Control Theory in Rheology

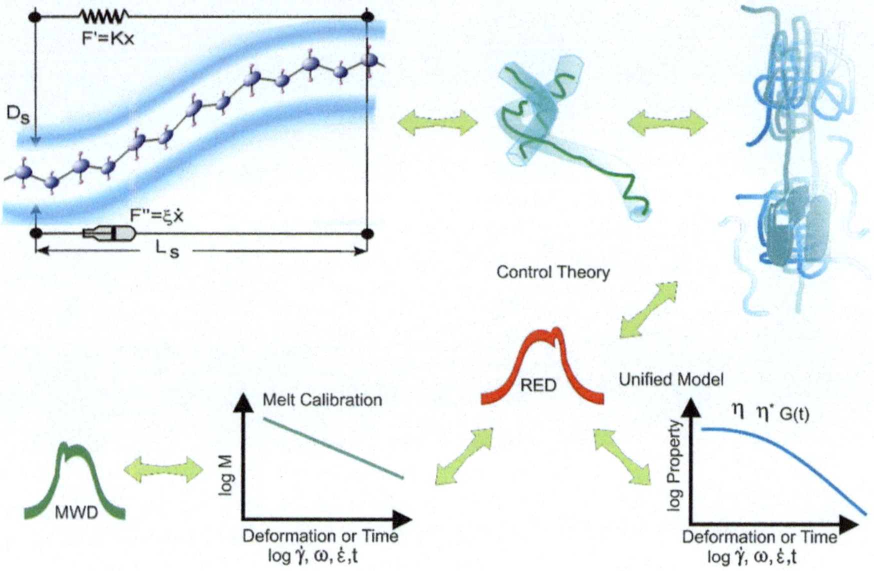
F'=Kx
Ds
F"=ξẋ
Ls
Control Theory
RED
Unified Model
Melt Calibration
log M
MWD
Deformation or Time
log γ̇, ω, ε̇, t
η η* G(t)
log Property
Deformation or Time
log γ̇, ω, ε̇, t

Tommi Borg

Control Theory in Rheology

An Introduction to Practical Applications

Springer

Tommi Borg
Evijärvi, Finland

ISBN 978-3-031-88247-0 ISBN 978-3-031-88248-7 (eBook)
https://doi.org/10.1007/978-3-031-88248-7

This Springer imprint is published by the registered company Springer Nature Switzerland AG
The registered company address is: Gewerbestrasse 11, 6330 Cham, Switzerland

Preface

This book provides a connection between theoretical rheology and industrial applications by introducing Control Theory (CT) and the linear Unified Model. This framework facilitates the modelling and analysis of various viscoelastic flows, as well as polymer and macromolecular structures.

The seed for this book was planted over 50 years ago when, as a young summer trainee at an extrusion plant, I made my first rheological calculations for engineers at the age of 17. At that time, available books focused on theoretical relaxation spectra and similar topics, offering little help for the practical problems encountered in industry. Fortunately, in recent decades, books aimed at practical engineering applications have emerged. These include empirical formulas, such as the Power Law for viscosity, as well as experimental equations for injection moulding, extrusion, and other fields, supported by numerous CAD-aided software tools.

Modern polymer studies and textbooks primarily emphasize polymer structures and measured properties, often without providing robust models for viscoelastic flows. This book offers an alternative rheological foundation, presenting a scientific and mathematical linear approach suitable for future textbooks and related research. It provides extensive insights and explanations of polymer flow and structural analysis of macromolecules.

Many useful viscoelastic constitutive equations are presented for analysing and modelling shear and complex flows, relaxation modulus and spectrum, elongation, transient viscosity, and for computing the Molecular Weight Distribution (MWD) from viscoelastic measurements. The book also covers temperature dependence and hemorheological analyses through CT. Furthermore, it offers deeper insights into the background of power-law theories and the Cox-Merz rule, concluding with comparisons to classical and Newtonian models.

Finally, I would like to extend my heartfelt thanks to all those who have contributed to making this book possible. I am grateful to my business customers and the workers at extrusion, injection moulding and polymerization plants, whose practical challenges inspired much of the content. I also deeply appreciate the anonymous referees of scientific journals, whose constructive criticism greatly improved the structure and clarity of the manuscripts.

Special thanks go to my co-author for the manuscripts, Prof. Dr. Esko J. Pääkkönen of Tampere University of Technology, whose tireless efforts and guidance greatly refined my academic writing. Our collaboration involved hundreds of emails, phone calls, and other communications, and his dedication to this shared endeavour has been invaluable.

Evijärvi, Finland
January 2025

Tommi Borg

Contents

About the Author

Tommi Borg It is often said that today's children, as young as five, are adept at using computers. My journey began similarly, but 65 years ago, at the dawn of the sixties. At that young age, I was captivated by the open-view viscoelastic paste and extrudates in my parents' factory. Over time, my understanding evolved, leading me to realize that the world of plastics is not solely governed by relaxation modulus and spectrum.

Since there were no local opportunities to study rheology at the time, I pursued studies in Computer Science, Physics, and Mathematics, writing my first trial program 50 years ago. Artificial Intelligence (AI) still relies on classical numerical methods as its basis. However, Unified Model formulas are essential for detecting MWD or structures, as these are highly sensitive to viscoelastic data errors and require precise input for accurate solutions.

Having spent my entire career in plastics conversion, design, process invention, and software development, and later working as a consultant for global companies, this book represents the culmination of my life's work.

Chapter 1
Theoretical and Practical Rheology

This book bridges the gap between theoretical rheology and industry applications by utilizing the widely adopted and scientifically validated Control Theory (CT) to analyse and model various types of viscoelastic flows and macromolecular structures.

In practical engineering, the design of machinery and equipment for polymers often relies on handbooks, respective textbooks, and numerous CAD-aided software tools based on empirical formulas.

The original theories of viscoelasticity in rheology, with their main principles developed over a century ago, are rooted in spring-and-dashpot elements. Furthermore, many books and studies in theoretical rheology attempt to dissect all the variables of a polymer or viscoelastic flow in accordance with the principles of sound physics practice. This leads to a quagmire and complicated formulas without practical use. And still not possible variables effected by material structure is included. In the CT principle, this is achieved in a novel way and applied to the linear *Unified Model*—referred to as the Characteristic Model in a published paper—for different viscoelastic flows.

1.1 The Origin of Classical Rheology

Viscoelasticity is most commonly described using the generalized Maxwell model (also known as the Maxwell-Wiechert model), which was first introduced in 1889. This constitutive model involves a parallel arrangement of spring-and-dashpot elements and incorporates the stress of the relaxation modulus to include the concept of relaxation time.

A comprehensive overview of the theory of linear viscoelasticity was succinctly summarized by Gross [1] in his work *Mathematical Structure of the Theories of Viscoelasticity*, published by Hermann & Cie, Paris, in 1953. Gross observed that the formalisms of **dielectric** and **viscoelastic** behaviour are analogical, drawing from

T. Borg, *Control Theory in Rheology*, https://doi.org/10.1007/978-3-031-88248-7_1

his experience as an electrical researcher. He presented many relations and formulas for different viscoelastic flows and authored a concise rheological book, which also served as an inspiration for this present book.

He termed his approach "dogmatic" because it was grounded on fundamental principles of dielectric formulas, enabling the theory to be presented in a compact and lucid manner. Unfortunately, Gross's book remained relatively obscure, overshadowed by the advancements in rheology highlighted by Ferry [2]. Additionally, numerous significant works and books have since adopted this phenomenological theory of linear viscoelastic behaviour as a foundational reference.

Esteemed figures such as Tschoegl [3], Dealy and Larson [4], Larson [5] and Graessley [6, 7], among others have advanced the continuum of Gross's "dogmatic" approach through well-documented works, contributing significantly to the body of knowledge reviewed by Gross. Unfortunately, many modern textbooks have forgotten even to refer him and words of analogy with dielectric behaviour. Secondary as original theory is very limited dealing relaxation phenomena having only narrow view of polymer flows, authors have widened subject with many artificial components.

An alternative approach for practical engineering applications involves using empirical formulas, such as the Power Law for viscosity, along with other experimental equations. Similarly, for extruder design, resources like handbooks by Michaeli [8] and Rauwendaal [9] are commonly used. For injection moulding modelling, the principles established by Kennedy and Zheng [10] are widely applied.

Modern polymer studies and textbooks focus primarily on polymer structures and measured properties. This book offers an alternative rheological foundation for extensive studies and explanations of polymer flow and structural analysis, presenting a scientific and mathematical linear approach suitable for modern textbooks and related research works.

1.2 Presented Rheology Principle is Based on Control Theory

The rheology principle presented here is rooted in Control Theory. We demonstrated, using RheoPower software [11], that macromolecules do not exhibit behaviour exactly akin to the classical rheology principles derived from dielectric phenomena. Originally, these principles described stress relaxation states for viscoelastic materials. Our proposed principle overcomes these limitations and is applicable to various flow types and conditions.

The development process began in the 1970s, drawing insights from practical manufacturing and the design of manufacturing equipment and tools for various industries. It underwent several step-by-step iterations, even with the assistance of early personal computers in the begin of eighties, before we realized that our model essentially aligned with the theoretically accepted Control Theory in the realm of science.

Maxwell was actually the first in 1868 to introduce a mathematical model for Control Theory and governors used with steam engines and other devices [12]. The viscoelastic Maxwell model, incorporating spring and dashpot components, was presented two decades later. Recently, many engineering and mathematical disciplines have adopted Control Theory, but, as far as we know, it has never been applied to rheology before.

When we attempted to publish peer-reviewed scientific papers in esteemed rheological journals, they consistently asked for a relation to classical rheology. Thus those published papers came complicated and non-chronological. As principle starts on a clean table and on new standpoint in rheology, developments cannot be evaluated by Occam's razor principle. Actually, only few material-dependent constants are used.

In summary, our work, based on Control Theory and conducted to the linear *Unified Model* for different viscoelastic flows, maintains a higher level of accuracy, flexibility, and notably, a more "dogmatic" constitutive approach compared to the classical presentation by Gross et al. later on, which stemmed from dielectric phenomena. However, our approach possesses a much more universally grounded physical foundation and enhanced flexibility derived from Control Theory, which is extensively used in various physics and engineering branches. In summary, we can refer to this as the new theory of the linear viscoelastic behaviour of macromolecules and other viscoelastic materials by Control Theory and Unified Model. One novel finding was that not all molecules impact simultaneously to the flow properties.

1.3 Main Lines for the Composition of the Book

To make the book understandable on its own, in the Chaps. 2 and 3 is presented fundamentals of distributions and Control theory. Chapters 4–13 are based on our peer-reviewed scientific papers in esteemed rheological journals [13–18], but the presentation is ordered in a more logical and chronological way. For practical applications steady shear viscosity is discussed in Chap. 4 and complex viscosity in Chap. 5. Relaxation modus in Chap. 6 and all those Chapters are mead as stand alone.

As the six published papers introduced over 60 new equations during the development of the principle, the presentation initially became non-sequential. Only minor modifications have been made to the nomenclature and the use of logarithms to standardize distributions on the same scale and ensure congruence.

Calculation examples via spreadsheet have been added to clarify the use of the formulas. Chapters 14–15 are already based on partly peer-reviewed scientific papers; however, we interrupted the publishing process [19–21].

Many figures and computed results have been omitted to keep the presentation compact, but necessary clarifications have been added. You can find more and original figures in our published papers.

In the first Chapters is shown accurate linear relations between different flows and rheological structure. At the later Chapters is discussed more theoretical background, conversion to chemical structure and computation results. Spring and dashpot constitutive models are used with temperature dependences at molar structure level in Chap. 12 and differences with Maxwell model is discussed in Chap. 16. Elsewhere, classical formulas are not referenced; instead, some experimental results are used.

The highly regarded term "Newtonian," in its various forms, is similarly mentioned towards the end, as it was not deemed necessary for the discussions surrounding the Control Theory (CT) presented in this book.

This book focuses exclusively on the relationships between polymer flows and response structures through the CT principle, echoing Albert Einstein's words, "Everything should be made as simple as possible, but not simpler." The CT and Unified Model relations provide a robust platform with precise viscoelastic relationships for further research into structures and complicated viscoelastic flows.

References

1. Gross, B. (1953). *Mathematical structure of the theories of viscoelasticity*, Hermann & Cie, Paris.
2. Ferry, J. D. (1980). *Viscoelastic properties of polymers* (3rd ed.). Wiley-Interscience.
3. Tschoegl, N. W. (1989). *The phenomenological theory of linear viscoelastic behavior: An introduction.* Springer.
4. Dealy, J. M., & Larson, R. G. (2006). *Structure and rheology of molten polymers*. Hanser Publishers.
5. Larson, R. G. (1999). *The structure and rheology of complex fluids*. Oxford University Press.
6. Graessley, W. W. (2003). *Polymeric liquids & networks structure and properties*. Garland Science.
7. Graessley, W. W. (2008). *Polymeric liquids & networks: Dynamics and rheology*. Taylor and Francis Grup.
8. Rauwendaal, C. (2014). *Polymer extrusion* (5th ed.). Carl Hanser Verlag.
9. Michaeli, W. (2003). *Extrusion dies for plastics and rubber* (3rd ed.). Hanser Publishers.
10. Kennedy, P., & Zheng, R. (2013). *Flow analysis of injection molds* (2nd ed.). Hanser Publishers.
11. https://www.tomcoat.com
12. Maxwell, J. C. (1868). On governors. *Proceedings of the Royal Society of London, 16*, 270–283.
13. Borg, T., & Pääkkönen, E. J. (2009). Linear viscoelastic models. Part I. Relaxation modulus and melt calibration. *Journal of Non-Newtonian Fluid Mechanics, 156*, 121–128.
14. Borg, T., & Pääkkönen, E. J. (2009). Linear viscoelastic models. Part II. Recovery of the molecular weight distribution using viscosity data. *Journal of Non-Newtonian Fluid Mechanics, 156*, 129–138.
15. Borg, T., & Pääkkönen, E. J. (2009). Linear viscoelastic models. Part III. Start-up and transient flow effects from the molecular weight distribution. *Journal of Non-Newtonian Fluid Mechanics, 159*, 17–25.
16. Borg, T., & Pääkkönen, E. J. (2010). Linear viscoelastic models. Part IV. From molecular dynamics to temperature and viscoelastic relations using control theory. *Journal of Non-Newtonian Fluid Mechanics, 165*, 24–31.
17. Borg, T., & Pääkkönen, E. J. (2011). Linear viscoelastic model for elongational viscosity by control theory. *Rheologica Acta, 51*, 371–384.
18. Borg, T., & Pääkkönen, E. J. (2015). Linear viscoelastic model for different flows based on control theory. *Applied Rheology, 25*, 1–10.

19. Borg, T., & Pääkkönen, E. J. (2013). Polymer structures and flows by new linear viscoelastic principle, non-published.
20. Borg, T., & Pääkkönen, E. J. (2015). A new computational rheological method for determining polymer structure from viscosity measurements, non-published.
21. Borg, T., & Pääkkönen E. J. (2022). A new computational hemorheological method for analyses, non-published.

Chapter 2
The Principle of Control Theory (CT)

This chapter provides some foundational theory behind the formulas presented in this book. However, for those focused on practical calculations, delving into this chapter is optional but recommended for a comprehensive understanding.

2.1 The Basement for Control Theory (CT)

Modern control theory is widely employed in various technologies such as principles in mathematics, nuclear physics, signal control systems, digital computers, aircraft design and processing industries. However, it has not been previously applied to viscoelasticity. To apply control theory to model the flows of real polymers [1], we first consider a system excited by a small stress induced by a small pulsed strain applied at time t_0. The pulse response, denoted as $y(t)$, is obtained from the impulse response, denoted as $h(t)$, and by sampling the active molecules' distribution, $w(t)$, over some time interval as follows:

$$y(t) = \int_{-\infty}^{\infty} w(\tau)h(t - \tau)\mathrm{d}\tau \tag{2.1}$$

Here, τ is the standard integration variable. The convolution equation is a well-known linear formula used in control theory to regulate and manage a variable or function within a closed system. Relations to molecular theories is discussed later in many Sections as 2.5 Molecular Model of Macromolecules, in the Chap. 4 and subsequent chapters.

T. Borg, *Control Theory in Rheology*, https://doi.org/10.1007/978-3-031-88248-7_2

2.2 Convolution Integral Example

Convolution is a mathematical operation that produces a new function derived from two given functions. This method is widely applied in image and signal processing, engineering, and physics. Although numerical computation of convolution is fundamental and straightforward for computers and software, performing it manually can be quite time-consuming.

The convolution of two functions, $w(t)$ and $h(t)$, is defined as:

$$(w * h)(t) = \int_{-\infty}^{\infty} w(\tau)h(t - \tau)\mathrm{d}\tau = y(t) \tag{2.2}$$

Let's consider a simplified "discrete convolution" as a numerical example. While this isn't a true continuous convolution, it serves as an illustrative representation of the concept. Consider the functions

$$f(t) = 2t + 1$$
$$g(t) = t^2 + 2$$

Evaluating these functions for $t = 1$ and $t = 2$, we get for $(f{*}g)(t)$:

$$f(1) \times g(1) + f(1) \times g(2) + f(2) \times g(1) + f(2) \times g(2) = 72$$

This showcase that performing convolution calculations manually can be cumbersome. In contrast, computers handle such computations with ease. In fact, our daily-use devices, such as PCs and smartphones, continually compute convolution integrals for tasks like image processing.

Luckily with polymers since some the greatest or longest chains have given their response, they are mainly ineffective for transitions of rheological effects. Still convolution is needed in the base and source formulas for modelling practical flows.

2.3 Example: Control Response of an Aircraft

Imagine an aircraft that has a specific response to a sudden change in the throttle setting, i.e., an impulse response. This impulse response characterizes how the aircraft's speed changes in response to a sudden, short-lived burst of throttle.

Let the impulse response $h(t)$ represent the aircraft's speed change in m/s over time (seconds) after a unit impulse (a very short, momentary burst of throttle). For simplicity, we can say:

$$h(t) = e^{-t}u(t)$$

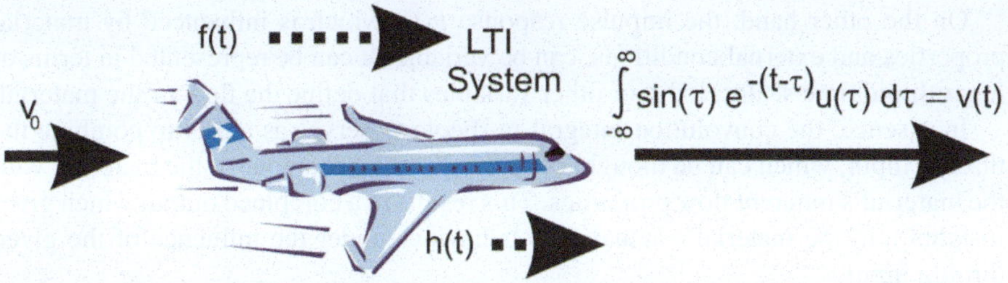

Fig. 2.1 Imagine an airplane's flight as a linear time-invariant (LTI) system at speed v_0. The system has new pilot's throttle input, $f(t)$, and the aircraft's speed response, $v(t)$. We'll convolve the throttle input function $f(t)$ with the impulse response $h(t)$ of the aircraft

where $u(t)$ is the unit step function.

Now, suppose the pilot's throttle input (the control input) over a period is described by:

$$f(t) = \sin(t)u(t)$$

To find the aircraft's speed response $v(t)$ over time to this throttle input, we'll convolve the throttle input function $f(t)$ with the impulse response $h(t)$:

$$v(t) = \int_{-\infty}^{\infty} \sin(\tau)e^{-(t-\tau)}u(\tau)d\tau$$

This convolution essentially combines the aircraft's inherent response to control changes with the specific control input given by the pilot, yielding the speed response over time (see Fig. 2.1).

In the real world, the convolution would give a more complicated result, and the impulse response of the aircraft would be based on empirical testing or high-fidelity simulations. But this example serves as a basic introduction to how control theory and the convolution integral can describe an aircraft's response to inputs.

2.4 Rheological Flows and the Convolution Integral

In rheological studies, understanding the flow behaviour of materials is of paramount importance. The convolution integral, which has been discussed in the context of other applications, finds its relevance here as well.

Imagine the throttle input function $f(t)$ as a representation of a fraction of the Rheologically Effective Distribution (RED) or Molecular Weight Distribution (MWD) functions. This throttle function acts as an external stimulus or input, much like the way an aeroplane's throttle would influence its velocity.

On the other hand, the impulse response $h(t)$, which is influenced by material properties and external conditions, can be variable. It can be represented in terms of logarithmic time scales, log t, or other variables that define the flow of the material.

In essence, the convolution integral in rheology serves as a tool to combine the throttle input, which can be thought of as the desired behaviour of the material, with the material's inherent flow properties. This results in a combined output which gives insights into the material's actual flow behaviour under the influence of the given throttle input.

This convolutional approach offers a holistic view, allowing us to understand how different throttle inputs or RED/MWD distributions affect the flow behaviour of the material, given its inherent properties and the external conditions it's subjected to. In summary, in the airplane example, the throttle acted as a simple impulse input function $f(t)$ analogous to the RED/MWD fraction. Conversely, the impulse response $h(t)$ represents a straightforward function on a logarithmic scale for the rheological CT formulas.

In the following sections, we will delve deeper into specific rheological flows and formulas, drawing parallels with the convolution integral, and demonstrating its applicability in predicting and analysing the flow behaviour of various materials.

2.5 Molecular Model of Macromolecules

A segment of a polymer chain is modelled under conditions of steady-state shear [2]. In practice, molecules exhibit branching and entanglements, which will be discussed later. The molecular chain vibrates within a statistical tube, and during shear, friction and stress forces come into play. This phenomenon can be visualized using the analogy of a mechanical spring and dashpot, as depicted in Fig. 2.2 shown.

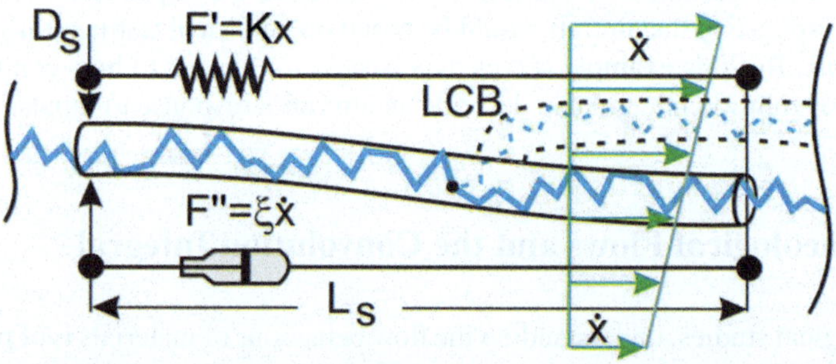

Fig. 2.2 Dimensions of an oriented unit segment within an imagined shear tube, which undergo changes at different temperatures. An oriented unit segment subjected to high shear is depicted with roughly scaled dimensions used for modelling. The presence of possible long-chain branching (LCB) results in an increase in the statistical unit segment length denoted as L_S, and elevated temperature causes an expansion in the diameter of the tube labelled as D_S. From Borg and Pääkkönen [2]

When we embark on the task of modelling flow by dissecting variables into elements, following classical physical procedures, we encounter an array of never-ending issues and intricate formulas. Fortunately, within Control Theory, we have the capability to analyse multiple active variables at various speeds, as previously discussed through the function $f(t)$. This methodology is employed to introduce the concept of the Rheologically Effective Distribution (RED), denoted as $w(t)$, which establishes connections among chain dimensions, forces, and, ultimately, the Molecular Weight Distribution (MWD) at the endpoint.

This connection is established through a process of Melt Calibration discussed in the Chap. 7, which will be introduced in more detail in the Sect. 12.2, “Relations to molecular dynamics”.

2.6 Conducted to the Unified Model for Different Viscoelastic Flows

During modelling real measured data were found that not all molecules impact simultaneously to the flow properties at new state or shear rate. This conducted development to the Unified Model still related to the CT discussed in the Chap. 4 and subsequent chapters.

2.7 Linear Principle Between Micro- and Macrostructure

Principle is applicable to the use of shear viscosity, complex viscosity, dynamic moduli, relaxation modulus, extension and other time-dependent models and flows at various temperatures. Given that CT and the formulas are linear, one can proceed in the transverse direction to derive Rheologically Effective Distribution (RED) from viscoelastic measurements and further MWD through Melt Calibration conversion.

Melt Calibration is the relation between time, rate or frequency and the molecular weight fractions. This procedure has similarities with the widely used universal calibration with dissolved polymers. The RED function can be regarded as a GPC trace with time-exclusion chromatography or SEC elugram function as a function of time (see Fig. 2.3).

As RED is according to viscoelastic measurements and properties, it can inform much more about polymer structure as MWD. Moreover, the measurements of MWD have been unrealistic during time for polyolefins as polydispersity value and average molecular weight has decreased 80% as discussed in the Sect. 3.8.

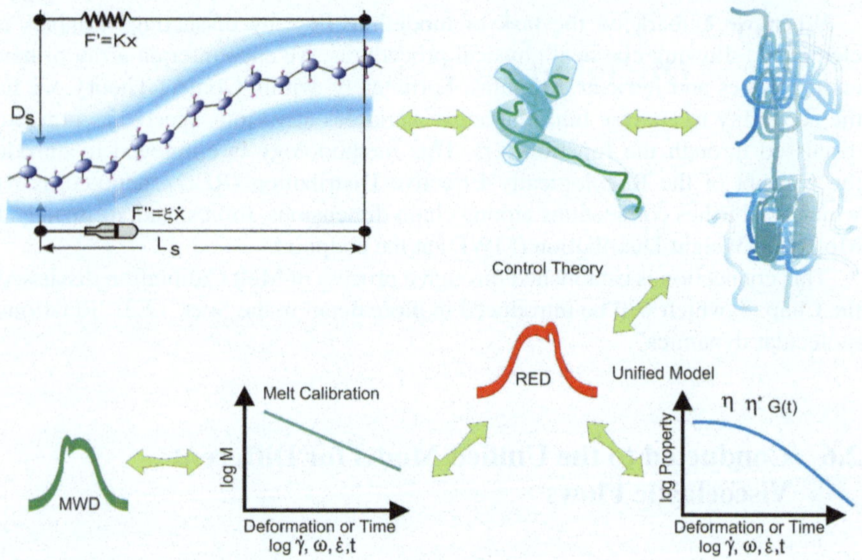

Fig. 2.3 Graphical demonstration of the system using Control Theory and the nanostructure model of a polymer melt. Relative diameters of the statistical tube for molecules of different sizes produce the accurate Rheologically Effective Distribution (RED), which is subsequently converted to the MWD through melt calibration. The signal bump is shown to observe and separate RED or MWD. *Unified Model* sets accurate linear relation between REDs and different type viscoelastic flows

References

1. Borg, T., & Pääkkönen, E. J. (2009). Linear viscoelastic models. Part I. Relaxation modulus and melt calibration. *Journal of Non-Newtonian Fluid Mechanics, 156*, 121–128.
2. Borg, T., & Pääkkönen, E. J. (2010). Linear viscoelastic models. Part IV. From molecular dynamics to temperature and viscoelastic relations using control theory. *Journal of Non-Newtonian Fluid Mechanics, 165*, 24–31.

Chapter 3
Basics of Distribution Functions

Distribution functions in polymers are discussed. For a more accurate modelling of phenomena in physics or materials, especially polymers and composite materials made of elements of varying types or sizes, an approach slightly more complex than mere algebraic base formulas is required. This involves considering the composition of various material fractions, known as Molecular Weight Distribution (MWD). To conduct rheological modelling and computations with Control Theory (CT) and Unified Models, it is necessary to utilize the Rheologically Effective Distribution (RED) related to MWD.

The macrostructure of synthetic polymers is inherently heterogeneous. The term 'polymer' implies a composition of varying molecular sizes and structures. Unlike the classical normal or Gaussian distribution often encountered in the sciences, the representation of molecular weight distribution (MWD) on a logarithmic scale may seem somewhat unclear. In this section, we will clarify and compare these different distribution patterns, highlighting their unique characteristics and implications. Experience has shown that what appears to be straightforward regarding distribution has led to misunderstandings and incorrect conclusions.

3.1 Non-logarithmic Standard Distribution

Initially, the classical non-logarithmic distribution and its corresponding equation are presented. In a normal distribution, the parameter μ represents the mean, while the parameter σ denotes its standard deviation. Figure 3.1 shows the distribution across two decades, although the standard widths for MWD and REDs typically span several more decades. Additionally, CT and Unified formulas operate on logarithmic scales.

$$f(x) = \frac{1}{\sigma\sqrt{2\pi}} e^{-\frac{1}{2}\left(\frac{x-\mu}{\sigma}\right)^2} \tag{3.1}$$

T. Borg, *Control Theory in Rheology*, https://doi.org/10.1007/978-3-031-88248-7_3

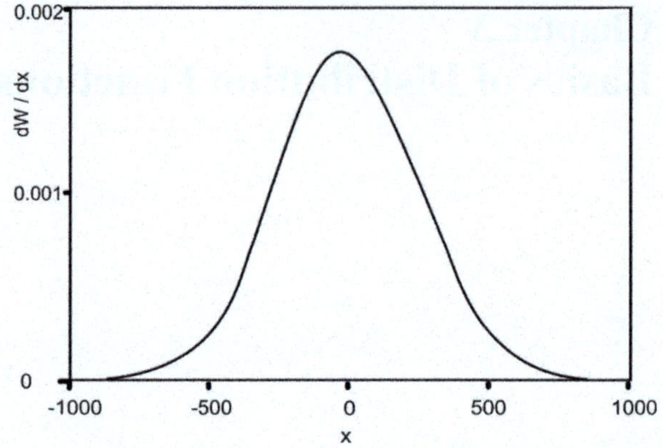

Fig. 3.1 Graphical Normal or Gaussian distribution with mean $\mu = -100$, variance $\sigma^2 = 225$. Differentially normalized distributions $f(x)/dx$ is good practice to make comparisons between distributions

The parameter μ is the mean or expectation of the distribution (and also its median and mode), while the parameter σ is its standard deviation. σ^2 is the variance.

These non-logarithmic distributions are not used with polymers as it is not at all uncommon to have significant contributions to the distribution that extend over three or four decades of molecular weight distribution. In such cases, it is much more convenient to express the weight distribution in terms of logarithmic increments in molecular weight and respectively REDs.

3.2 Natural and Common Logarithms

In scientific mathematical analyses, natural logarithms are typically used for convenience, denoted as log x or sometimes ln x. Similarly, the first Molecular Weight Distributions (MWD) represented using $w(\log M)$ actually were often displayed on a natural logarithm scale $w(\ln M)$. Nowadays, the common logarithm scale, $w(\log_{10} M)$, is more frequently used for polymer distributions. We standardize all distributions generated on spreadsheets for this book to the same scale and ensure congruence by using this common logarithm, marked $w(\log M)$. However, our MWDs published in journals are on a natural logarithmic scale, and the scale of some other published MWDs is unknown.

The difference in fraction values, with normalization of the natural logarithm d ln M, results in fraction values being over two times less for f ($f/\ln 10 \approx 0.434f$), and conversely, the common logarithm d log M gives f ($f/\log 10 \approx 2.30f$). For used formulas different scales of distributions has no influence for results as they are normalized during computation.

This book generally uses the common logarithmic form, and necessary exceptions in the natural logarithmic form are marked by ln notation.

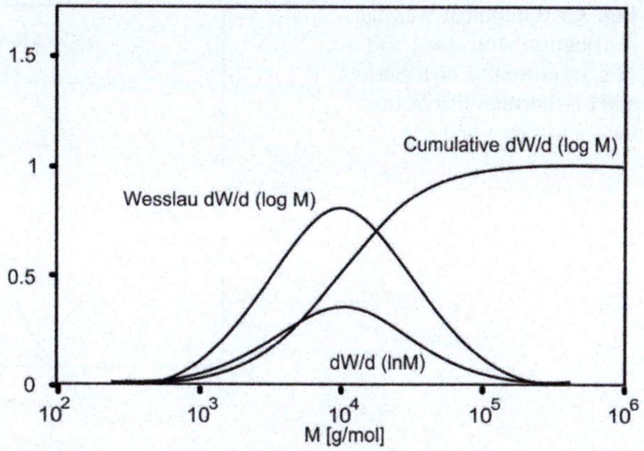

Fig. 3.2 Graphical Wesslau distribution with $\mu = 4$ and $\sigma = 0.5$ for MWD chart. Difference with fraction values with normalization of natural logarithm d ln M gives fraction values over two times less for fraction f ($f/\ln 10 \approx 0.434f$). Cumulative distribution approaches value 1

3.3 Wesslau Distribution

With polymer and other viscoelastic materials, it is convenient to use logarithmic scales. Respective for normal distribution known also as Gaussian is the logarithmic normal distribution function also known as Wesslau distribution shown on Fig. 3.2 on the logarithmic molecule weight scale, which form is typical of addition and Ziegler–Natta catalysed systems.

$$f(\log x) = \frac{1}{\sigma\sqrt{2\pi}} e^{-\frac{1}{2}\left(\frac{(\log x - \mu)}{\sigma}\right)^2} \tag{3.2}$$

We use mainly this generated RED distribution $f(\log x)$ in the modelling viscoelastic flows on examples. Some REDs used are transformed from true MWDs.

As you can obtain, only differences with normal distribution are the fact that variable x is in the logarithmic form $\log x$. As most presented formulas based on the CT principle, distributions are on the logarithmic scale. Equation (3.2) has many names, one is the Wesslau distribution [1], but Watterson suggested the name the logarithmic normal distribution function [2]. For us it is standard normal distribution on the logarithmic scale and world and is called here Wesslau distribution for shortness and to separate from log-normal distribution presented in the next section.

3.4 Log-Normal Distribution

Log-normal distribution Eq. (3.3) is different matter as the Wesslau distribution Eq. (3.2).

$$f(\log x) = \frac{1}{x\sigma\sqrt{2\pi}} e^{-\frac{(\log x - \mu)^2}{2\sigma^2}} \tag{3.3}$$

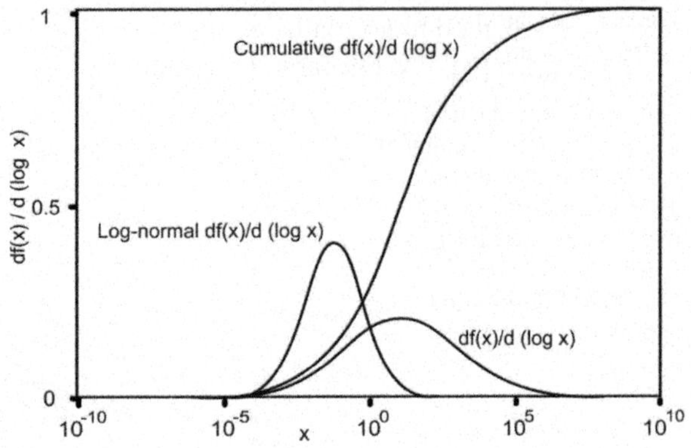

Fig. 3.3 Graphical Wesslau distribution with $\mu = 1$ and $\sigma = 2$, its cumulate distribution and log-normal distribution the same $\mu = 1$ and $\sigma = 2$

The log-normal distribution, sometimes called the Galton distribution, is a probability distribution whose logarithm has a normal distribution. Graphical Wesslau distribution, its cumulate distribution and log-normal distribution are shown on Fig. 3.3. As a function of flow rate or time simulating standard RED function.

All figures and charts as a function of different rates or time on the logarithmic scale are presented on the range from 10^{-10} up to 10^{10} due to used arrays or column ranges are the same in the example sections. The ends of distributions need space. If needed for illustration reason, then is used narrower ranges.

3.5 Normalization of Differential Molecular Weight Distribution (MWD)

RED and MWD are on the statistical standpoint probability density distributions. For molecular weight distribution (MWD) the respective formula is $w(\log M)$, where M is molecular weight g/mol and w is distribution. For computations even on spreadsheet have to do discretization to distributions to get numerical mode of all information.

But polymer data is converted into a differential form, which describes how the quantity of molecular fractions changes as a function of molecular weight. With MWD is usually expressed in the form $\frac{dW}{d\log M}$, where W is the weight percentage, and M is the molecular weight.

The aim of normalization is to adjust the graph so that its area corresponds to 100%, facilitating the comparison of different samples. This is done by integrating the differential graph (i.e., calculating the area under the graph) and dividing each data point's value by this total area. Mathematically, this can be expressed as:

$$\text{Normalized value} = \frac{\text{Differential value}}{\text{Integrated area}}$$

This process ensures that the total of all values on the graph corresponds to 100%.

MWDs are good to normalize into a differential form to make those comparable each other. On the MWD charts vertical fraction scale is marked $\mathrm{d}W/\mathrm{d}(\log M)$ but older documents the fraction values are not even shown. Please obtain, that sometimes earlier used with GPC/SEC charts for normalization natural logarithm giving fraction values $\mathrm{d}W$ over two times less ($1/\ln 10 \approx 0.434$).

3.6 The Difference of Fraction dW/d(log M) and w(log M)

Some words about the distinction between $\mathrm{d}W/\mathrm{d}(\log M)$ and $w(\log M)$. The relative weight fraction of polymer fractions within a certain range of molecular weight is marked $w(\log M)$. It is not a differential quantity, but rather a discrete or cumulative distribution that shows the weight percentage for a specific molecular weight range. This is typically presented as a histogram or bar chart, where each bar represents the weight fraction of a particular molecular weight category.

On the contrary $\mathrm{d}W/\mathrm{d}(\log M)$ is a differential representation that indicates how the weight fraction of polymer fractions changes with respect to the logarithm of the molecular weight. Mathematically, it is the derivative of the distribution with respect to the logarithm of the molecular weight. It describes how much the weight fraction changes with a small change in molecular weight. This is typically represented as a curve on a graph, where each point on the curve represents the differential weight fraction at a specific molecular weight point.

3.7 Normalization of Differential Rheologically Effective RED

Respectively with REDs weight fractions are relating true physical weights. We use same procedure as for differential MWD and as rheologically effect fractions are also true as virtual weights, we use general form for RED distributions $\mathrm{d}W/\mathrm{d}(\log x)$ on chart and graphic. Thus vertical fraction scale is marked with REDs in the form $\mathrm{d}W/\mathrm{d}(\log x)$. Variable x for horizontal scale is $\log \dot{\gamma}$ for shear rate flows, $\log \omega$ for complex shear or frequency rate ω, $\log t$ as a function of time and $\log \dot{\varepsilon}$ for elongation rates.

The CT method is using mainly relative values in the logarithmic space with few absolute values. We show some base formulas on the logarithmic scale. Non-normalized distribution function $\tilde{w}(\log x)$ is marked with the tilde mark "~" on the top.

In the numerical form with software and spreadsheet is used discrete form of distribution. This can be achieved by summarising all fractions $\tilde{w}(\log x_i)$ together to get sum value c as follows.

$$c = \sum_{i=1}^{n} \tilde{w}(\log x_i)(\log x_i - \log x_{i-1}) \tag{3.4}$$

to get normalised differential distribution as follows.

$$dW/d(\log x) = \tilde{w}(\log x)/c \tag{3.5}$$

In the charts is used for RED $w(\log x)$ fractions $dW/d(\log x)$ or other variables as shear rates or time. Although there is not real molecule fractions in the RED, the used form indicates relation and inverse form to the MWD.

3.8 Normalization of the Cumulative Distribution for RED

The procedure for RED $w(\log x)$ to get normalized cumulative distribution $\mathrm{W}(\log x)$ is very essential modelling viscosity flows according to Control Theory. Normalization of integral Eq. (3.6)

$$\tilde{W}(\log x) = \int_{-\infty}^{\infty} w(\log x)\,\mathrm{d}\log x \tag{3.6}$$

is done when Eq. (3.7) is true:

$$\lim_{x \to \infty} W_x(\log x) = 1 \tag{3.7}$$

Normalization procedure of cumulative RED $w(\log\ x)$ can be written also in shortly form shown Eq. (3.8), but we present procedure for spreadsheet with other important matters and the used row scale for presentations here.

$$\int_{-\infty}^{\infty} w(\log x)\,\mathrm{d}\log x = 1 \tag{3.8}$$

Normalization can be represented as scaling the cumulative distribution function $W(x)$ such that $W(x_{\max}) = 1$ or 100%, where $x_{\max}$ is the maximum variable x in the distribution.

The cumulative distribution function on the logarithmic scale is the probability that the variable takes a value less than or equal to log x. In the CT principle we use here general rate or time variable x, which can be $\dot{\gamma}$ for shear rate flows, ω for complex shear or frequency rate, t for as a function of time and $\dot{\varepsilon}$ for elongation rates.

We write the integral for cumulative distribution function $\tilde{W}(\log x)$ as follows:

$$\tilde{W}(\log x) = \int_{-\infty}^{\log x} w(\log \psi)\mathrm{d}\log \psi \tag{3.9}$$

For a discrete cumulative distribution, the formula can be expressed as

$$\tilde{W}(\log x) = \sum_{i=1}^{x} f(\log x_i)(\log x_i - \log x_{i-1}) \tag{3.10}$$

We get maximum value of the discrete cumulative distribution $\tilde{W}(\log x)$ by setting x value approaching infinite value or simply setting $x_{\max}$ value of to the Eq. (3.10) and get sum value c as follows.

$$c = \tilde{W}(\log x_{\max}) \tag{3.11}$$

Finally normalization by Eq. (3.11)

$$W(\log x) = \frac{\tilde{F}(\log x)}{c} \tag{3.12}$$

As summary we get relations shown in the Eq. (3.12)

$$P'(\log x) = P'F(\log x) = P' \int_{-\infty}^{\log x} w'(\log \psi)\mathrm{d}\log \psi \tag{3.13}$$

Similar discrete cumulative distribution normalization must do for respective viscous distribution and integral shown Eq. (3.13), where discrete variable x_i shown in Eq. (3.12) has now form $\log\frac{x_i}{R}$.

$$P''(\log x) = P''F(\log x) = P'' \int_{-\infty}^{\log x} w''\left(\log\frac{\psi}{R}\right)\mathrm{d}\log \psi \tag{3.14}$$

where Rouse constant, R, the distance for the same point for the effective Rouse relaxation and entanglement relaxations discussed in Sect. 4.5.

With software and spreadsheet normalization of cumulative distributions is easy work to do by selecting the value of end cell or array for the sum value c.

3.9 Unreliable MWD Results by GPC/SEC During Decades for LDPE

Using weight average molecular weight $\overline{M}_W$ and polydispersity value *MwR* of LDPE mislead researcher decades in developing methods and software detecting MWD from rheological measurements. Here is marked $\overline{M}_W$ as average molecular weight *Mw* to get unform marking with software outputs. Well analysed commercial polymer BASF Lupolen 1840H and its ancestor grades during years as "IUPAC A" and "Melt I", originally named by Meissner [3] and Laun [4], are characterized by *Mw* and *MwR* values, which has lost their number about 80% as can see on Table 3.1. Conversion factors *Mf* and *Hf* for complex viscosity are also shown.

Values for *Mw* and *MwR* are left only 20% from the first measurements. Respectively conversion factor to simulate GPC/SEC results are set by RheoPower software for polymer structure *Hf* and between scales by *MwR* values to model according polymer flow measurements. Before year 2005 also some tiny modifications were made to the computation of RheoPower arrays making earlier *Hf* values are not tightly comparative.

Year 1975 MWD *Mw* and *MwR* data is by Meissner [3], year 2001 all data is from personal communication from Basf, which used for MWD form is done by RheoPower software, year 2005 some new computation result and year 2022 *Mw* and *MwR* by Poh et al. [5].

We used complex viscosity data measured by Basf year 2001 to compute RED. This RED data we imported here to standard spreadsheet and simulated MWDs results according to years. Values by principle are easily scaled according to chemical methods by using the polymer structural *Mf* value to set *Mw*. Respectively conversion factor *Hf* between frequency rate and molar weight scales is set by *MwR* value. Relaxation modulus and spectrum simulations were done by year 1975 data and with other simulations is told the year of MWD data source.

We show respective MWDs during on Fig. 3.4, the first one had names as "Melt I" and IUPAC A, but they are surely from different polymerization batches.

Not vain is pointed out that during working with REDs and MWD distributions is essential to confirm the right function. Those are easily blended which each other.

Although the GPC/SEC methods are based on extensive experimental data and has proven useful in many applications, its accuracy and applicability can vary depending on the chemical properties of the polymers and their behaviour in solution. Therefore,

Table 3.1 Values for *Mw* and *MwR* over the years

Year	Mw	MwR	Mf	Hf
1975	467935	24.6	≈37800	≈2.05
2001	234689	14.1	37800	2.5
2005	234689	14.1	26700	2.72
2022	91955	4.53	23000	3.7

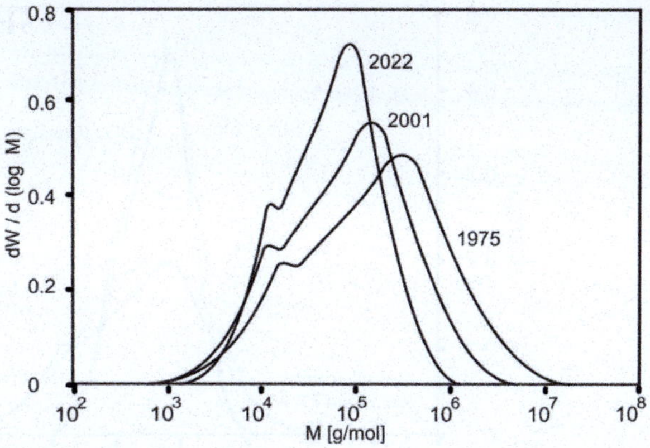

Fig. 3.4 Lupolen 1840H LDPE and its ancestors Melt I" and IUPAC A MWDs at different decades. Small pumps on the left side are coming from derivation of viscosity data. They are left without smoothing as they are excellent sign to obtain, if distribution is RED or MWD

it is important to understand its limitations and apply it critically in the analysis of different types of samples.

MWD topics are discussed in the Melt Calibration Chap. 7. Formulas for average molecular weight *Mw* and polydispersity index *MwR* are presented there.

3.10 Different REDs for Different Type Flows

Originally REDs were detected and computed from polymer flow measurements. Unfortunately, to measure some flows is difficult in practice. In this case is possible to use the earlier obtained RED for the polymer grade, as computed MWD of material and real structure is constant. We need only to use different *Mf* and *Hf* values by Melt Calibration method for our temporally generated or true MWD for new type flow. This transformed new RED of the different type flow is base of modelling. Transformed REDs on spreadsheet is shown on Fig. 3.5.

For a while the base shape of RED is similar at different flows inside the accuracy. Only differences are with "polydispersity" and "average mean" of RED using correspondence meanings of MWD.

In the upcoming Chap. 4, we will display 1840H LDPE and Wesslau REDs for shear rate and capillary flows. Complex viscosity is discussed in Chap. 5, while the relaxation modulus and spectrum are covered in Chap. 6.

A crucial point is that, although MWDs of 1840H LDPE may vary over the years, the RED remains constant in relation to the viscoelastic properties. Here are used only two distinct REDs.

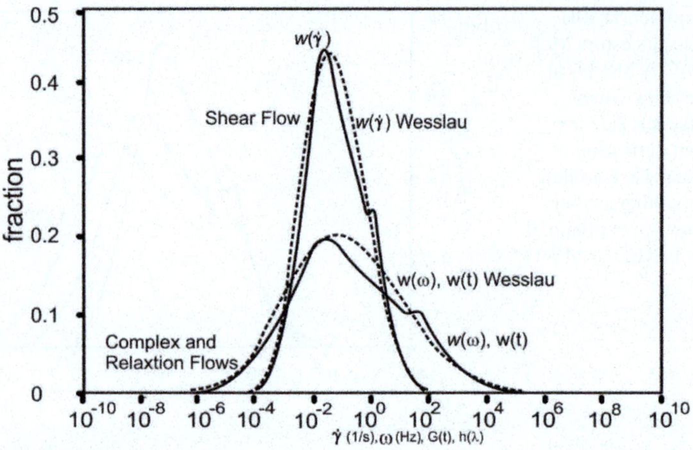

Fig. 3.5 1840H LDPE polymer REDs for different flow types. We can obtain that steady shear data scans much wider RED thus polymer structure compared complex shear rate or relaxation modulus. This is important information for developing MWD analyses by CT. For shear rate the respective Wesslau distribution $\mu = -1.4$ and $\sigma = 0.9$. Complex shear rate or relaxation modulus thus polymer structure Wesslau distribution $\mu = -0.9$ and $\sigma = 1.92$

3.11 Spreadsheet Examples

We use spreadsheets to demonstrate calculations as they are surely well known for everyone. User can then use other mathematical tools which are more familiar and in the daily use. There are hundred mathematical software and coding languages beside dozens different spreadsheet platforms. But the presentation on a spreadsheet clarifies some special points of the used CT principle and discrete software arrays:

All main computations are done on the logarithmic scales and word.

Presenting modelling distributions on the logarithmic scales.

Descripting concretively the source of viscous and elastic effects and give understanding of respective distributions (REDs).

3.11.1 Number of Spreadsheet Rows

Number of the used index for data elements (table rows, array index with software arrays) on column cells is 200. On a top row is name of variable or distribution. First variable is set on the row 10 and last one 210. This amount is enough for informative purposes. Once you have made the first spreadsheet calculation, the other base spreadsheets for different flows are close similar and you can use create a copy of spreadsheet command for new type flow.

On software can use number for indexes as 200, 4000, 40,000 or more depending on the purpose of computing.

Logarithmic rate variable x is on 0.1 steps from log(−10) to log(+10) or on a column −10, −9.9, −9.8….. done by simple addition −10 + 0.1 formula to the cells downwards.

Respectively logarithmic molar mass M is from log(0) to log(+10) or on a column 0, 0.05, 0.1….. done by simple addition 0 + 0.05.

3.11.2 Spreadsheet for Generating Distributions

We show pre-configured spreadsheet for calculations and in Fig. 3.7. At first a cut of top rows for a figure (see Fig. 3.6).

Please note that all spreadsheets shown here are used with the MS Excel 365 product, utilizing standard settings for the English language. Be aware that your regional format settings, including those for functions, decimal separators, array separators, and list separators, may differ. Another a lot of used spreadsheet is LibreOffice Calc.

	C	D	E	F	G	H	I	J	K	L
3	RED Wesslau									
4										
5	μ	1								
6	σ	2								
7										
8			10.00	1.00	10.00	91.00	10.00	7.08	0.71	10.00
9	x	log x	Wesslau f(log x)	f(log x)*d log ›	df()/d log x	cumul. df(x)/d log x	log Norm	Log-normal df(x)/d log x		Log-normal df(x)/d log x
10	0	-10	5.38488E-08	7.0805E-09	7.081E-09	7.0805E-10	5.38E-08	1.05941E-17	2.5155E-18	1.49555E-17
11	0	-9.9	7.0805E-08	7.0805E-09	7.081E-08	7.78855E-09	7.08E-08	2.51545E-17	2.5155E-18	3.55103E-17

Fig. 3.6 Presents a top section of an MS Excel 365 spreadsheet used for modelling distributions. Cells outlined with thick borders and containing bold numbers are designated for input data. Cells framed by thinner borders represent output data or are intended solely for verification purposes. You can make a chart selecting the whole area except non-log most left column x and then delete unnecessary chart lines. On the Excel select Insert/Charts/Scatter and there select "Scatter with Smooth Lines" or "Scatter with Smooth Lines and Markers". Then you can adjust ranges and outlook

	C	D	E	F	G	H	I	J	K	L
3	RED									
4										
5	μ	1								
6	σ	2								
7										
8			`=+SUM(E10:E210)`	`=+SUM(F10:F210)`	`=+SUM(G10:G210)`	`=+SUM(H10:H210)`	`=+SUM(I10:I210)`	`=+SUM(J10:J210)`	`=+SUM(K10:K210)`	`=+SUM(L10:L210)`
9	x	log x	Wesslau f(log x)	f(log x)*d log x	df'()/d log x	cumul. df(x)/d log x	log Norm	Log-normal df(x)/d log x		Log-normal df(x)/d log x
10	`=10^D10`	-10	`=1/($D$6*SQRT(2*PI()))*EXP(-1/2*(((LOG10(C10)-D$5)/$D$6))^2)`	`=+F11`	`=+F10`	`=+G10/G$8`	`=NORM.DIST($D10;$D$5;$D$6;FALSE)`	`=1/($C10*$D$6*SQRT(2*PI()))*EXP(-((LOG10(C10)-D`	`=+K11`	`=+J10/K$8`
11	`=10^D11`	`=+D10+0.1`	`=1/($D$6*SQRT(2*PI()))*EXP(-1/2*(((LOG10(C11)-D$5)/$D$6))^2)`	`=+(D11-D10)*E11`	`=+E11/F$8`	`=+H10+F11`	`=NORM.DIST($D11;$D$5;$D$6;FALSE)`	`=1/($C11*$D$6*SQRT(2*PI()))*EXP(-((LOG10(C11)-D`	`=+(D11-D10)*J11`	`=+J11/K$8`

Fig. 3.7 Used formulas of the Fig. 3.7 in "formula auditing mode"

3.11.3 Formulas on Spreadsheet

By selecting "formula auditing mode" on the Formula tab in Excel ribbon tab you can see formulas. In the "Formula Auditing" group click on the Show Formulas option. All the Formulas in the active worksheet will be displayed, so that at once you will get to know which cells contain formulas and what the formulas are.

At first rows is Wesslau distribution, the Log-Normal according to Eq. 3.3 and Log-Normal Excel function on the Column "I" is not suggested to use.

In the first row 10 can be initial values, so that row 11 is the first real formula row. On the cell C10 is the first value for log x = −10.

When you get ready formulas on the row 11, then by copy (Ctrl C) whole row and paste (Ctrl V) shortcut commands you can fill easily the whole range with 200 rows. There are also available some tricks to fill quickly, but this is out of cope.

3.11.4 List of Formulas in the Cells

Input values:

D5 1 (μ is the mean)
D6 2 (σ is standard deviation)
D10 −10 (initial log rate for rows and arrays)

Functions:

F8 =SUM(F10:F210)
K8 =SUM(K10:K210)
C11 =10^D11
D11 =+D10+0.1
E11 =1/($C$6*SQRT(2*PI()))*EXP(−1/2*((LOG10(C11) – $C$5)/$C$6)^2)
F11 = (D11 – D10)*E11
G11 =+E11/F$8
H11 =+H10 + F11
I11 =NORM.DIST($D11;$D$5;$D$6;FALSE)
J11 =1/($C11*$D$6*SQRT(2*PI()))*EXP(−((LOG10(C11) – D$6)^2)/(2*D$5^2))
K11 =+ (D11 – D10)*J11
L11 =+J11/K$8

In this case was not needed to use normalization of cumulative distribution to get highest value 1 by procedure explained in the Sect. 3.7.

Normal or Gaussian distribution function NORM.DIST is only for comparison. As we are modelling mainly on logarithmic scales, is better to use Wesslau function formula or generated from flow measurements or Melt Calibration from MWD.

References

1. Wesslau, H. (1956). *Makromolekulare Chemie, 20*, 111.
2. Watterson, J. G. (1971). The logarithmic normal distribution function. *Journal of Macromolecular Science: Part A—Chemistry, 5*(5), 1007–1009.
3. Meissner, J. (1975). Basic parameters, melt rheology, processing and end-use properties of three similar low-density polyethylene samples. *Pure and Applied Chemistry, 42*, 551–662.
4. Laun, H. M. (1982). Elastic properties of polyethylene melts at high shear rates with respect to extrusion. *Rheologica Acta, 21*, 464–469.
5. Poh, L., Wu, Q., Chen, Y., & Narimissa, E. (2022). Characterization of industrial low-density polyethylene: A thermal, dynamic mechanical, and rheological investigation. *Rheologica Acta, 61*, 701–720.

Chapter 4
Steady Shear Viscosity

In industrial applications, the flow of steady shear melt holds paramount importance. Following a theoretical introduction to Control Theory, the practical *Unified Model* Eq. (4.6) [1] is presented, illustrating that not all molecules contribute to the flow properties simultaneously. This model was referred to as the Characteristic Model in our published paper. A more thorough explanation of the background used from the standpoint of the relaxation modulus is presented in Chap. 6.

4.1 Steady Shear Rate Viscosity

Traditionally, steady simple shear flow $\eta(\dot{\gamma})$ is referred to as the "viscometric flow." Utilizing linear control theory combined with RED/MWD, we demonstrate that this viscometric function actually represents linear viscoelasticity. Additionally, the time-dependent transient shear viscosity in practical applications will be discussed in Chap. 10. In the examples, we use 1840H LDPE MWD described in Fig. 3.4, a descendant of classical IUPAC A LDPE, with data measured by the Weissenberg Rheogoniometer (WRG) by Meissner [2], and "Melt I" measured by Laun with a sandwich rheometer (SSR) [3], which is in principle the optimal on the standpoint of Unified Model. Some high shear rate data are measured using a capillary tube, necessitating Bagley corrections (see Fig. 4.1).

T. Borg, *Control Theory in Rheology*, https://doi.org/10.1007/978-3-031-88248-7_4

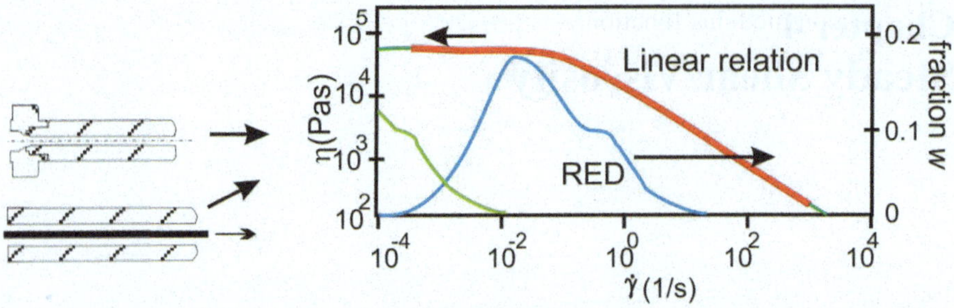

Fig. 4.1 Figure demonstrates how a steady shear rate for a sample can be achieved by moving the central plate between two plates, as shown in the lower left, or it can also be measured using a capillary tube at high rates. A linear mathematical relationship has been identified between the Rheologically Effective Distribution (RED) and the shear viscosity flow curve $\eta(\dot{\gamma})$. The thicker red coloured section of the measured viscosity data aligns accurately with the modelled viscosity

4.2 Starting Our Model from Known Principles

Published molecular theories in rheology are mainly using relaxation phenomena and we start by similar way [4]. They are based on independent chain response (elastic dumb-bell, Rouse, Zimm) or pseudo-independent chain response (unmodified Doi-Edwards) can be used for predicting of the MWD according to Graessley [5] in *Polymeric Liquids and Networks: Dynamics and Rheology* (2008). For such systems, equations for viscosity, recoverable compliance and relaxation time in systems of arbitrary polydispersity can be obtained from the stress relaxation modulus, expressed as the sum of independent contributions. Thus $G(t) = \sum w_i G_i(t)$, in which w_i is the weight fraction of component i in the mixture, and $G_i(t)$ is the relaxation modulus for the monodisperse component.

We continue to develop the above principle further by converting weight fraction $w_i(M)$ as a function of molecular weight M to the function of time t or *effective fraction* $w_i'(t) = w_i(M)$ with the found relation between t and M discussed in Chap. 7. A similar conversion, but in the opposite direction, is widely used in wet chemical methods such as gel-permeation chromatography (GPC) tracing with time- or size-exclusion chromatography (SEC) for detecting the MWD.

Knowledge of the factors that influence the effective fraction, $w_i(t)$, is not essential, such as different types of chain dynamics, molecular friction and elasticity, reptation, primitive path fluctuations, constraint release, and other entangled and disentangled chain dynamics.

Chain types in the effective fraction $w_i(t)$ can conform to any linear short- or long-chain branched (SCB and LCB, respectively) structures, combs, networks, H-shapes with multiple arms, or any combinations of complete molecules or their segments. The final structure, molecular weight or construction of the statistical $w_i(t)$ fraction of a single molecule or groups of molecules is not important at this point—only its effects on viscoelasticity. Chapter 7 discusses the real sizes of the fractions. Now we get relaxation modulus functional as presented by Anderssen and Loy [6] for $G(t)$ as follows: $G(t) = \sum w_i(t)G_i(t)$.

We introduce the *rheologically effective distribution* (RED), $w(t)$, and impulse response $h(t)$ after induced stress at time t_0. Relaxation modulus $G_i(t)$ in has relation to a scaled product of $PG_0h(t_i)$, in which constants G_0 is zero modulus and P is acting as a scaling factor. Thus we can rewrite for $G(t)$ as

$$G(t) = G_0 P \int_{-\infty}^{t} w(\tau)h(t-\tau)\mathrm{d}\tau \tag{4.1}$$

which can be explained and developed further by the control theory with the explanation starting from the basis. We excite the system with a small stress induced by a small pulsed strain that is applied at time t_0. Pulse response $y(t)$ is obtained from impulse response $h(t)$ and by sampling the active molecules in distribution $w(t)$ between some time interval:

$$y(t) = \int_{-\infty}^{t} w(\tau)h(t-\tau)\mathrm{d}\tau \tag{4.2}$$

This equation is the familiar linear formula used in control theory, which is known as a novel principle to adjust and rule one variable or function in a closed system and now in our case for MWD. More about relaxation modulus is discussed in the Chap. 6.

4.3 Preliminary for MWDs and REDs

Before going to the main subject of the chapter we present a demonstrative example about the difference between MWD and RED. The organizers of the "Second Workshop on Inverse Problems in Rheology and Related Experimental Techniques", held in Germany during May 9–11, 2001, arranged a public benchmark test for participants to carry out and present their own results. PS samples were so-called blind data, in that their origin and manufacturer were unknown. The data for two samples were received at a workshop, as explained below, and were simply marked "gstern_2080" and "gstern". We extracted PS data from G' and G'' master curves to develop η^* as arranged by the organizers and used RheoPower software to get thick lines to Fig. 4.2. These narrow—but still bimodal—MWD results are wholly Analytical according to Unified Model: chemically measured MWDs are indicated by dashed lines as given in the MWD forms distributed by the workshop organizers.

Computed by RheoPower software we obtained following respective REDs shown in Fig. 4.3. We used Melt Calibration principle discussed in Chap. 7 with constants for Polystyrene PS $Hf = 4$ and polymer structure value $Mf = 13{,}600$ g/mol.

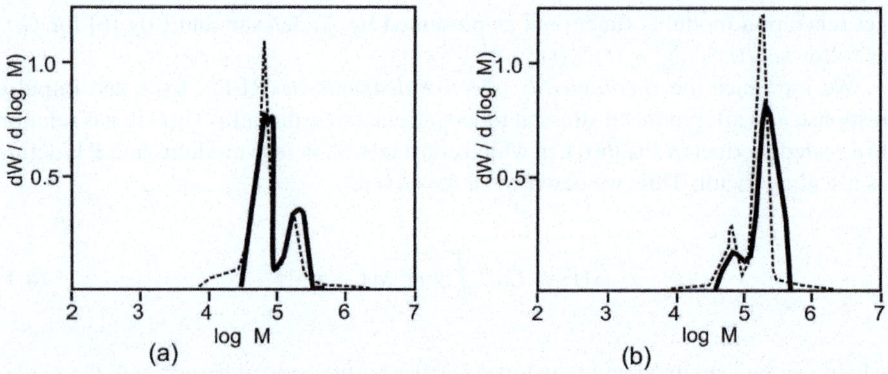

Fig. 4.2 PS benchmark test for workshop participants to carry out and present their own results for MWDs from blind data: **a** "gstern_2080", **b** "gstern". Dotted lines are measured by GPC/SEC shown later on. Borg and Pääkkönen [1]

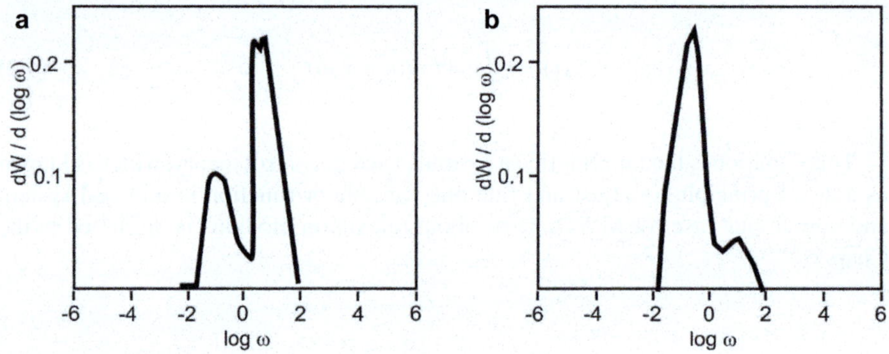

Fig. 4.3 Shows respective REDs for PS sample in same order: **a** "gstern_2080", **b** "gstern". Please obtain that REDs are mirror figures to the respective MWDs

4.4 Shear Viscosity Model by Control Theory

We develop steady shear viscosity formula based on the control theory principle as in Sect. 4.2 for relaxation modulus. We use MWD as a function of shear rate by *effective fraction* $w_i'(\dot{\gamma}) = w_i(M)$ from function of molecular weight M in which w_i is the weight fraction of component i in the mixture. Now the development procedure can be started from the principle $\eta(\dot{\gamma}) = \sum w_i(M)\eta_i(\dot{\gamma})$ as sum of independent contributions to get complex viscosity functional for $\eta(\dot{\gamma}) = \sum w_i(\dot{\gamma})\eta_i(\dot{\gamma})$.

The basic idea is to model steady impulses and the summed stress resulting from chain dynamics between molecules of different molecular sizes at a steady shear rate $\dot{\gamma}$. With the normalized rheologically effective distribution (RED), $w(\dot{\gamma})$, and the impulse response, $h(\dot{\gamma})$, we obtain system response $y(\dot{\gamma})$ according to control theory as follows:

$$y(\dot{\gamma}) = \int_{-\infty}^{\dot{\gamma}} w(\psi)h(\dot{\gamma} - \psi)\mathrm{d}\psi \qquad (4.3)$$

where ψ is the standard integration variable for shear rate $\dot{\gamma}$. This equation is a linear formula used in control theory to adjust and control one variable or function in a closed system. We applied this formula to determine the $w(\log \dot{\gamma})$ distribution function (RED) and also molecular weight distribution $w(\log M)$ (MWD) shown in the Chap. 7.

We rewrite all variables and functions in Eq. (4.3) on a logarithmic scale. The logarithmic convolution (also known as the scale convolution) can be related to ordinary convolution by taking the logarithm of $\dot{\gamma}$. We can, therefore, write $h(\dot{\gamma}) = \dot{\gamma}$ and $\mathrm{d}\dot{\gamma}$ as $\mathrm{d}(\log \dot{\gamma})$. System response viscosity $\eta(\log \dot{\gamma})$ is a normalized viscosity $\log\frac{\eta(\dot{\gamma})}{\eta_0}$ on the logarithmic scale at $\dot{\gamma}_0$. The zero viscosity at rate 1/s, η_0, is obtained by fitting to experimental measurements. We then obtain the general shear viscosity function $\eta(\dot{\gamma})$:

$$\log\frac{\eta(\dot{\gamma})}{\eta_0} = -P\int_{-\infty}^{\log \dot{\gamma}} w(\log \psi)\log\frac{\dot{\gamma}}{\psi}\mathrm{d}\log \psi \qquad (4.4)$$

where response $\log \dot{\gamma} - \log \psi = \log\frac{\dot{\gamma}}{\psi}$ and where ψ is the standard integration variable. Normalization of the cumulative distribution $\int_{-\infty}^{\infty} w(\log \dot{\gamma})\, \mathrm{d}\log \dot{\gamma}$ have to do and procedure is explained in the Sect. 3.7.

4.5 Dividing to Elastic and Viscous Components

Polymer structure studies are mostly done for relaxation processes or by complex moduli measurements, and we use this as an introduction to explaining the dividing procedure. Real polymer relaxation occurs in two phases: first, Rouse relaxation [7] and later in the second phase with reptation, collisions, and entanglement in longitudinal modes, all of which we refer to here as entanglement relaxation. Molecular entanglements will open and relax after deformation at times $t > 0.001$ s, where most $G(t)$ observations are made or shear viscosity $\eta(\dot{\gamma})$ measurements. An explanation of the procedure is discussed in Chap. 6, which focuses on the relaxation modulus and more familiar relaxation phenomena. We have the same $w(\log \dot{\gamma})$ RED related to MWD acting for both Rouse relaxation and later for entanglement relaxation in a similar way for changes of shear rates, where steady-state shear relaxation for active chains relaxation occurs during ceasing shear rate or retardation during rising rate discussed in Chap. 10.

We use the entanglement distribution in a common logarithmic scale $w(\log \dot{\gamma})$ for the earlier Rouse distribution, $w(\log \dot{\gamma} - \log R) = \mathrm{w}(\log \dot{\gamma}/R)$, by standard time shifting. This is achieved simply by copying the function $w(\log \dot{\gamma})$ after dividing $\dot{\gamma}$ by the ratio, R, of the shear rate ranges. The distance for the same point for the effective Rouse relaxation and shear rate $\dot{\gamma}_R$ and entanglement relaxations and shear rate $\dot{\gamma}_e$ changes, and the respective distribution, w(log $\dot{\gamma}$), is $R = \frac{\dot{\gamma}_e}{\dot{\gamma}_R}$. Rouse relaxation occurs for $t < 0.001$ s. We convert the shear rate formula of Eq. (4.4) to the entanglement and Rousean relaxation described by Eq. (4.5) and impulse response. In practice, the same distribution, $w(\log \dot{\gamma})$, is copied on the shear rate scale, and we obtain the following complete shear rate formula where ψ is the standard integration variable:

$$\log \frac{\eta(\dot{\gamma})}{\eta_0} = -\int_{-\infty}^{\log \dot{\gamma}} \left(P' w'(\log \psi) + P'' w'' \left(\log \frac{\psi}{R} \right) \right) \log \frac{\dot{\gamma}}{\psi} \mathrm{d} \log \psi \tag{4.5}$$

where the scalar values are $0 < P' < 40$ for entanglement and $0 < P'' < 40$ for the Rouse relaxation ranges. The entanglement value P' is related to the number of effective entangled molecules in the distribution or respective, while the Rouse value P'' is related to the number of effective molecules with Rouse relaxations and viscous effects.

In the Eq. (4.5) the elastic component is $P'w'(\log \dot{\gamma})$ and effective viscous is $P''w''(\log \dot{\gamma})$. These cumulative integrals must be normalised before using Eq. (4.5) explained in the Sect. 3.7 also for spreadsheet calculations.

It is good practise to normalize MWD $\breve{w}(\log M)$ distribution, but RED $\breve{w}(\log t)$ after conversion must always normalized. This keep all constants and variables comparative and relative to each other.

The values of P' and P'' are not ad hoc but are developed by the software during the procedure of fitting the shear viscosity and $w(\log \dot{\gamma})$ distribution. They are mainly depending on the polymer chemical structure. We call this as Analytical Model related directly to the CT principle.

4.6 From Analytical Model to Practical Unified Model

Originally the CT principle was developed for detecting MWD from viscoelastic measurements as shown in Chap. 9. Due to linear principle model relations offered conversions between polymer structure and flows to both directions. Unfortunately convolution integrals with a kernel, as $\log \frac{\dot{\gamma}}{\psi}$ in Eq. (4.5), are a severely ill-posed problem to get RED function and further MWD.

To solve this problem Unified Model was developed from Analytical and deeper background of procedure is discussed in the Chap. 6. In Sect. 5.1, a sampling band width T is introduced for complex shear viscosity as a solution within the Analytical Model.

In the oscillating rheometers real shear rates for molecules varies a lot and frequency ω is a kind virtual integral of all shar rates at ω frequency. Secondary found with complex viscosity that Unified Model gives much more accurate results. Now was found that with steady shear viscosity the difference between models was dramatic as shown in the Sect. 11.7. Convolution integral includes all molecules to all changes of different shear rates for computing new shear viscosity value, which is not the case.

4.7 Unified Model for Shear Viscosity

During modelling were found, not all molecules are active to changes of shear rates and to the shear viscosity output of Eq. (4.5). Above, we showed an Analytical mode of CT, but we found some small differences between measurements and modelled data with complex viscosity and now greater with steady shear viscosity. Also, we need later on to get MWD by solving distribution $w(\log\dot{\gamma})$ out. Functions with a logarithmic kernel, as $\log\frac{\dot{\gamma}}{\psi}$ in convolution integral Eq. (4.5), are a severely ill-posed problem reported by Bruckner and Cheng [8]. Therefore, we present an alternative solution method, which also models more accurate shear viscosity $\eta(\dot{\gamma})$. We call this model as the *Unified Model*, and this principle is valid in the next sections for complex, transient, and elongation viscosity and relaxation modulus, where is discussed more about the background of model.

We realized that we do not count and include all molecules at some shear rate point. Thus, the generated formula for the above shear rate viscosity Eq. (4.5), where is omitted molecules and chains not active anymore for changes at some new rate:

$$\log\frac{\eta(\dot{\gamma})}{\eta_0} = -\log\frac{\dot{\gamma}}{\dot{\gamma}_c}\int_{-\infty}^{\log\dot{\gamma}}\left(P'w'(\log\psi) + P''w''\left(\log\frac{\psi}{R}\right)\right)\mathrm{d}\log\psi \tag{4.6}$$

We set the characteristic shear rate $\dot{\gamma}_c$ constant to the former kernel $\log\frac{\dot{\gamma}}{\dot{\gamma}_c}$, which is, moreover, normally constant 1/s during computing. Background for this is discussed in Sect. 6.3. The values for constants P' and P'' are obtained through a feedback procedure during computation or are known beforehand for a specific polymer type. Values for polymer structure constants for elastic P' and viscous P'' normally vary from 0.1 up to 0.9. These values and R are found during fitting the model to measurements to get the best fit, but R is normally constant with standard polymers. Normalization of the cumulative integral distribution $\int_{-\infty}^{\infty}\tilde{w}(\log\dot{\gamma})\,\mathrm{d}\log\dot{\gamma}$ is recommended, and the procedure is explained in Chap. 3.

This RED is a fingerprint of material and is related to the structure and polydispersity via the MWD. A novel idea is to use the normalized $\log\frac{\eta(\dot{\gamma})}{\eta_0}$ viscosity and cumulative $w(\log\dot{\gamma})$ distribution, for which the lack of absolute values does not cause difficulties. We have by converting Eq. (4.6) a usable formula for viscosity $\eta(\dot{\gamma})$ as a function of shear rate:

$$\eta(\dot{\gamma}) = \eta_0\, 10^{-\log\frac{\dot{\gamma}}{\dot{\gamma}_c} \int_{-\infty}^{\log\dot{\gamma}} \left(P' w'(\log\psi) + P'' w''\left(\log\frac{\psi}{R}\right)\right) d\log\psi} \tag{4.7}$$

We use the common logarithm for the left side of Eq. (4.6) $\log\frac{\eta(\dot{\gamma})}{\eta_0}$, to derive the power law formula directly. Distribution integrals are presented in common logarithmic form as well, facilitating the determination of polymer structure constants for elastic P' and viscous P'' that are comparable to each other.

Not doubt real shear viscosity may be totally different at lower over measured $\dot{\gamma} < 0.0001$/s and higher rates $\dot{\gamma} > 10{,}000$/s, but modelling spreadsheet and software needs wider data or array ranges for REDs. What is presented at final output is totally different matter, but here we present wide rate scales in figures.

In Fig. 4.4 is shown computation results for shear viscosity by 1840H LDPE polymer. Shear viscosity measurements by Meissner and Laun were copied semi-manually and the %RMSE error between modelled and measured data according to Eq. 9.7 is obtained 0.034%. Semi-manually means that zoomed measured data points from articles and fitted on the modelled viscosity curve as close and between datapoints. The form of used MWD was computed former in Chap. 5 by complex viscosity measurements, but the converted width for RED (as polydispersity with MWD) is found by fitting by *Hf* and average value of RED by *Mf* values to get best fit viscosity for modelling on spreadsheet related to MWD.

In the Sect. 13.7 is discussed that manually copied data analysed by %RMSE value viscosity fits done by Cross, Bersted and Neural Net models are in the range 1%. When we look original scattered datapoints, somewhere over 2% result was

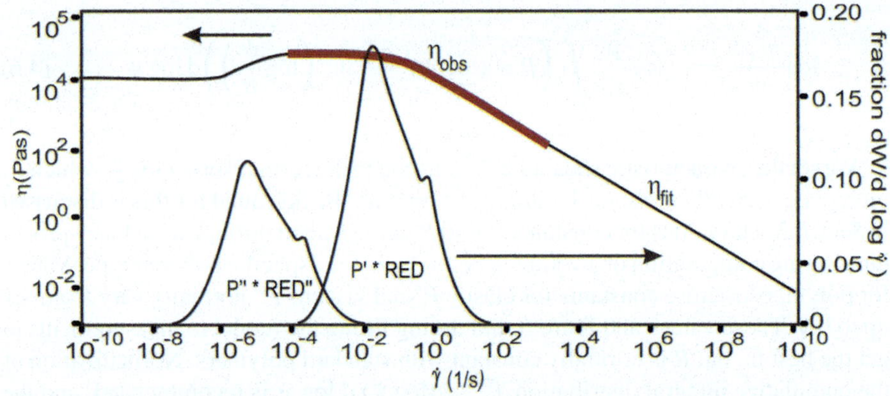

Fig. 4.4 A wide viscosity flow curve is simulated using a spreadsheet, with an example explained at the end of this chapter. RED is computed from complex viscosity measurements to obtain MWD based on information from the year 2022, with $Mf = 7000$ g/mol and $Hf = 1.7$, related to MWD. The elastic value $P' = 0.40$ and viscous $P'' = 0.22$ are used, which closely fits the shear viscosity at 150 °C by copied semi-manually from measurements by Meissner and Laun, as shown in the thick segment of the curve. In modelling, it is also necessary to use unrealistically low and high shear rates to prevent range limits from corrupting the results. When presenting viscosity results that include these ultra rate ranges elsewhere, it's essential to apply common sense

expected. By Unified Model is possible to get over hundred times smaller error with accurate measurements, as discussed later.

MWD is based on computation from complex viscosity measurements year 2005 [1], which form is converted according to measurement to the form shown in Fig. 3. 4 version year 2022 by procedures shown in Chap. 8. The width for RED (as polydispersity with MWD) is found by using $Hf = 1.7$ and average value of RED by setting $Mf = 7.000$ g/mol to get best fit viscosity for modelling on spreadsheet.

In our modelling, we also have to consider unrealistically low and high shear rates, which are impossible to verify through measurements. At these extreme rates, the actual viscosity likely diverges due to the influence of dynamic factors. Specifically, at very low rates, the elasticity factor P' is absent, and the viscosity may behave as modelled. However, should we manage to obtain measurements at these ultra rates, both our understanding and the model will improve.

4.8 Normalization of the Cumulative Elastic and Viscous Integrals

It is good practice to normalize MWD and RED distributions as discussed in Sect. 3. 5. However, cumulative sum integrals must be normalized for the CT formulas, with the procedure outlined in Sect. 3.8. Procedure was originally developed for computing MWD from viscosity data, where this normalization was essential also to get constants P' and P'' as a result of viscosity flow curves fits. Moreover, found later that polymer structure value P and function $P(\log \dot{\gamma})$ relates to the Power Law many other polymer properties.

General normalization of the cumulative distribution for RED $w(\log \dot{\gamma})$ can be written in shortly form shown Eq. (4.8) as follows and as explained in the Sect. 3.8:

$$\int_{-\infty}^{\infty} w(\log \dot{\gamma})\, \mathrm{d} \log \dot{\gamma} = 1 \tag{4.8}$$

4.9 Polymer Structure Functions

We present procedure in the Sect. 4.12 also for spreadsheet calculations. Then we can multiply normalized cumulative distributions with elasticity P' and P'' obtained by fitting according to measurements or gained by other way. For simplicity, we write Eq. (4.7) as an exponent without relation and originally kernel of convolution integral $-\log \frac{\dot{\gamma}}{\dot{\gamma}_c}$ and set up a very practical polymer structure function $P(\log \dot{\gamma})$

$$P(\log\dot{\gamma}) = \int_{-\infty}^{\log\dot{\gamma}} \left(P'w'(\log\psi) + P''w''\left(\log\frac{\psi}{R}\right)\right)\mathrm{d}\log\psi \qquad (4.9)$$

where components are:

$$P'(\log\dot{\gamma}) = P' \int_{-\infty}^{\log\dot{\gamma}} w'(\log\psi)\mathrm{d}\log\psi \qquad (4.10)$$

$$P''(\log\dot{\gamma}) = P'' \int_{-\infty}^{\log\dot{\gamma}} w''\left(\log\frac{\psi}{R}\right)\mathrm{d}\log\psi \qquad (4.11)$$

Figure 4.5 displays the component values of the Unified Model. The polymer structure function $P(\log\dot{\gamma})$ is the sum of the elastic $P'(\log\dot{\gamma})$ and viscous $P''(\log\dot{\gamma})$ cumulative integral components for LDPE H1840 at 150 °C. Measured viscosity or other viscoelastic data is shown across only a few decades, while MWD data spans closer to ten decades of the variable. However, in modelling according to the Unified Model, we utilize twenty decades. This approach is necessitated by the fact that a complete distribution or cumulative integral function is required for the formulas, even though the viscosity data at the low and high ends may be unrealistic.

Although absolute value level of rheologically effective distributions (REDs) $w'(\log\dot{\gamma})$ and $w''(\log\dot{\gamma})$ have no effect thanks to normalization of cumulative integral, a good habit is to use normalized before use in the CT principle. RED can be in the form $w(\log\dot{\gamma})$, $w(\log\omega)$, $w(\log t)$ or $w(\log\dot{\varepsilon})$. But respective integral functions

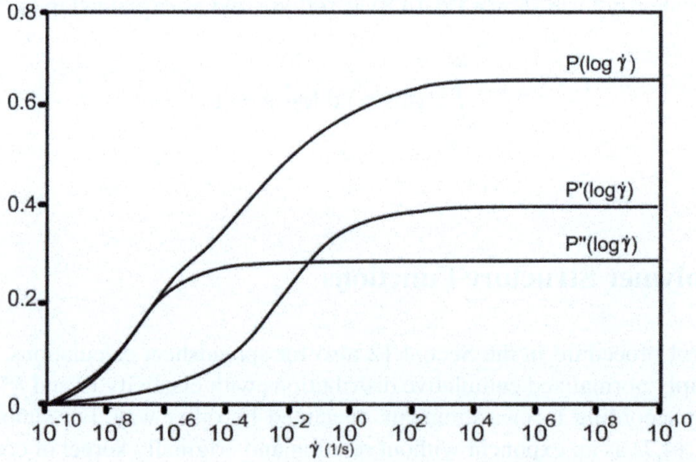

Fig. 4.5 Elastic $P'(\log\dot{\gamma})$, viscous $P''(\log\dot{\gamma})$ cumulative integral values of components and polymer structure function $P(\log\dot{\gamma})$ for LDPE H1840 by Eq. 4.12

$P'(\log \dot{\gamma})$ and $P''(\log \dot{\gamma})$ and as functions of ω, t and $\dot{\varepsilon}$ must be scaled similar way before multiplying by constants P' and P''.

According to the used principle, $P(\log \dot{\gamma})$ indicates the influence of effective viscous $w''(\log \dot{\gamma})$ and also changes in the elastic $w'(\log \dot{\gamma})$ distribution REDs. Besides incorporating the chain structure, it relates the orientation and degree of entanglement as functions of the shear rate.

It is much more convenient to use the short form of the equations. We write polymer structure function $P(\dot{\gamma})$ with non-logarithmic variable as follows:

$$P(\dot{\gamma}) = P'(\dot{\gamma}) + P''(\dot{\gamma}) \tag{4.12}$$

This gives us a simple power-form formula to Eq. (4.7) included relation $-\log\frac{\dot{\gamma}}{\dot{\gamma}_c}$ and approximates the true molecular orientation level and state.

$$\eta(\dot{\gamma}) = \eta_0 10^{-\log\frac{\dot{\gamma}}{\dot{\gamma}_c}P(\dot{\gamma})} = \eta_0\left(10^{\log\frac{\dot{\gamma}}{\dot{\gamma}_c}}\right)^{-P(\dot{\gamma})} \tag{4.13}$$

As we used common logarithm, we can write directly Eq. (4.14).

$$\eta(\dot{\gamma}) = \eta_0\left(\frac{\dot{\gamma}}{\dot{\gamma}_c}\right)^{-P(\dot{\gamma})} \tag{4.14}$$

4.10 Power Law

The power law (also called the Ostwald-de-Waele power law, which was first reported in 1925), usually written as $\eta = k(\dot{\gamma})^{n-1}$ for viscosity, is the simplified form of Eq. (4.14). Setting the characteristic shear rate $\dot{\gamma}_c = 1/\mathrm{s}$ in $\left(\frac{\dot{\gamma}}{\dot{\gamma}_c}\right)$ yields the classical power law for the shear viscosity as a function of shear rate:

$$\eta(\dot{\gamma}) = \eta_0|\dot{\gamma}|^{-P(\dot{\gamma})} \tag{4.15}$$

It's worth noting that η_0 differs from k in the power law$\eta = k(\dot{\gamma})^{n-1}$. The power law index n should be $P(\dot{\gamma}) = (1 - n)$ at higher shear rates. According to the background of the used model, P indicates and relates to the orientation, degree of disentanglement, and shear history of chains, which is an important property in CAE/CAD/CFD. The same procedure can be applied to determine the complex viscosity or relaxation modulus by CT principle.

4.11 Partitioning into Elastic and Viscous Components

We make partitioning shear viscosity into elastic σ' and viscous σ'' components analogically to dynamic measurements. Elastic and viscous components are impossible to measure with shear rate flow. Although we have a formula for viscosity in two components, we cannot use these components solely; we must also consider the influence of the other component. Anyway we use analogically similar model as for dynamic viscoelastic measurements discussed in the next Chap. 5. There we successfully succeed model partitioning by CT according to measurements of complex modulus G^*. As many want to study and get idea of viscoelasticity of material by storage G' and loss G'' moduli, we show respective formulas and figure for shear rate.

Dynamic viscoelastic measurements are done in *vibratory conditions,* what a steady shear is not. Thus, we model shear stress $\sigma(\dot{\gamma})$ as a function of shear rate as $\sigma = \eta\dot{\gamma}$. Also, we make partitioning into elastic σ' and viscous σ'' components analogically to dynamic measurements as the source is the same elastic RED and viscous RED$''$ distributions fitted on the shear flow (see Fig. 4.6).

$$\sigma(\dot{\gamma}) = \dot{\gamma}\,\eta_0\,10^{\left(-P'\log\left(\frac{\dot{\gamma}}{\dot{\gamma}_c}\right)\int_{-\infty}^{\log\dot{\gamma}} w'(\log\psi)\,\mathrm{d}\log\psi - P''\log\left(\frac{\dot{\gamma}}{\dot{\gamma}_c}\right)\int_{-\infty}^{\log\dot{\gamma}} w''\left(\log\frac{\psi}{R}\right)\mathrm{d}\log\psi\right)} \tag{4.16}$$

The exponent in Eq. (4.7), presented previously as $-\log\left(\frac{\dot{\gamma}}{\dot{\gamma}_c}\right)\left(P'(\dot{\gamma}) + P''(\dot{\gamma})\right)$ comprises two components of functions rewritten in Eq. (4.16) for shear stress

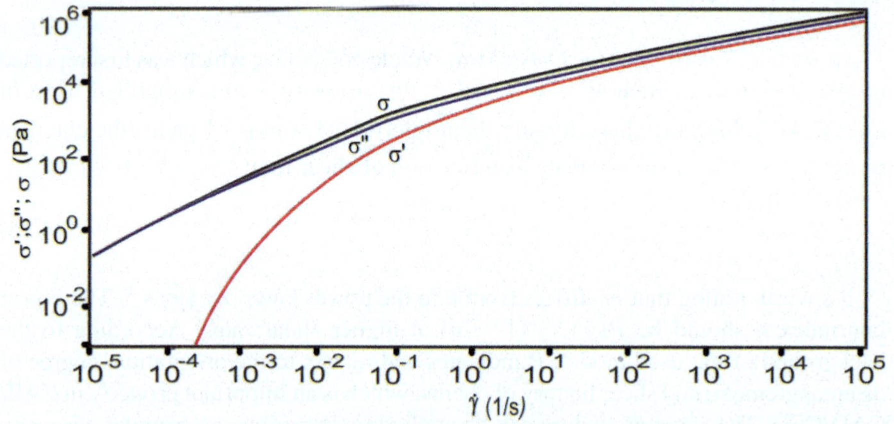

Fig. 4.6 Modelled storage σ', loss σ'' and σ stresses for 1840H LDPE (a modern version of IUPAC A) at 150 °C. There is no proof, that σ' and σ'' are right, although with complex viscosity modelled results were rather accurate with measurements. They are also more at sight separable as P' and P'' have close equal values with complex viscosity and moduli

$\sigma(\dot{\gamma})$. This formula can be converted in the exponential function form: $\sigma(\dot{\gamma}) = \dot{\gamma}\, \eta_0\, 10^{\log\left(\frac{\dot{\gamma}}{\dot{\gamma}_c}\right)(-P'(\dot{\gamma})-P''(\dot{\gamma}))} = \dot{\gamma}\, \eta_0\, e^{2.3\,\log\left(\frac{\dot{\gamma}}{\dot{\gamma}_c}\right)(-P'(\dot{\gamma})-P''(\dot{\gamma}))}$.

Exponential function without exponent part 2.3 $\log\left(\frac{\dot{\gamma}}{\dot{\gamma}_c}\right)$ can be expressed as $e^{-P'(\dot{\gamma})-P''(\dot{\gamma})}$, where functions $P'(\dot{\gamma})$ and $P''(\dot{\gamma})$ are given in Eqs. (4.9) and (4.10).

Obtained formula has similarities with complex notation. Thus we define exponential of complex variable z as $e^z = e^{-P'(\dot{\gamma})-iP''(\dot{\gamma})}$ related to complex modulus topic in Chap. 5.

We then partition Eq. (4.16) into components, where the same variable, 2.3 $\log\left(\frac{\dot{\gamma}}{\dot{\gamma}_c}\right)$ are divided between the numerator and denominator. The elastic stress σ' and viscous stress σ'' are defined as:

$$\sigma'(\dot{\gamma}) = \sigma(\dot{\gamma})\frac{P'(\dot{\gamma})}{\sqrt{P'^2(\dot{\gamma}) + P''^2(\dot{\gamma})}} \tag{4.17}$$

$$\sigma''(\dot{\gamma}) = \sigma(\dot{\gamma})\frac{P''(\dot{\gamma})}{\sqrt{P'^2(\dot{\gamma}) + P''^2(\dot{\gamma})}} \tag{4.18}$$

A complex number is nothing more than an ordered pair of two real numbers [9] and these functions are also real-valued. Similar partitioning is applied to the complex modulus in Chap. 5, whose results are comparable with those obtained through classical methods and direct measurements by an oscillating rheometer.

4.12 Converting by Melt Calibration Equations Between Scales

Converting the normalized rheologically effective distribution RED or $w(\log \dot{\gamma})$ to molecular weight distribution MWD $\tilde{w}(\log M)$ is discussed in the Melt Calibration Chap. 7, but we already present main formulas. For shear viscosity $\dot{\gamma}$, conversion equations are as follows:

$$\tilde{w}(\log M) = w(\log \dot{\gamma})\Big|_{\dot{\gamma}=\dot{\gamma}_1\left(\frac{Mf}{M}\right)^{Hf}} \tag{4.19}$$

$$\tilde{w}(\log \dot{\gamma}) = w(\log M)\Big|_{M=Mf\left(\frac{\dot{\gamma}}{\dot{\gamma}_1}\right)^{-\frac{1}{Hf}}} \tag{4.20}$$

The Mf is the polymer structural value that depends on the molecular structure and weight. Hf is the conversion factor between scales and is usually 2–4 with different flows, and Mf takes values in the range of 23,000–37800 g/mol for PE depending of the year measured MWD by GPC/SEC. For steady shear rate respectively $Mf =$

7000 g/mol and $Hf = 1.7$ gives same 1840H MWD measured year 2022 by GPC/SEC [10].

A good practise is the normalization of differential distributions according to procedure descripted in the Chap. 3 to get comparative $\mathrm{d}W/\mathrm{d}(\log M)$ and $\mathrm{d}W/\mathrm{d}(\log \dot{\gamma})$ for charts and figures.

4.13 Practical Examples

To approximate the wide function for shear viscosity, $(\dot{\gamma})$, we use a normalized Wesslau distribution for $w(\log \dot{\gamma})$ for RED. We then fit the appropriate form and position of the distribution using constants η_0, P', and P'' in the Sect. 3.11 "Spreadsheet Examples."

We use spreadsheets to demonstrate calculations as they are surely well known for everyone. User can then use other mathematical tools which are in the use. There are hundred mathematical software and coding languages beside dozens different spreadsheet platforms. All main computations are done on the common logarithmic scales and word. Discretization of distribution functions is convenient to use common logarithm base.

4.14 Example on a Spreadsheet

We model the true viscosity measurements of 1840H LDPE using artificial Wesslau distribution for RED, as shown in Fig. 3.5, with closely matching average molecular weight and polydispersity values to those of 1840H LDPE. Undoubtedly, measured MWD for 1840H LDPE can be utilized, but this data sample would require 200 rows for display in subsequent spreadsheets.

For Wesslau distribution is found the best fit viscosity by values $\mu = -1.4$ and $\sigma = 0.9$ to those modelled from Meissner and Laun's measurements. Additionally, the elastic value $P' = 0.40$ and viscous $P'' = 0.22$ were maintained consistently. The results are displayed in Fig. 4.7.

We now assume that the modelled viscosity data in Fig. 4.4 represent accurate measurements to simulate the effects of different variables on the results and the %RMSE error. With the same settings, we achieve a %RMSE of 0.27% for rates within $10^{-4}/\mathrm{s} < \dot{\gamma} < 10^3/\mathrm{s}$. Visually, we cannot discern any error between measured and obtained viscosity in this case, modelled using 1840H LDPE.

During the actual computation of MWD by software, the modified RED corrects errors in modelled viscosity according to measurements, thus reducing %RMSE to below 0.001, as demonstrated in Sect. 5.5, Chaps. 9 and 13.

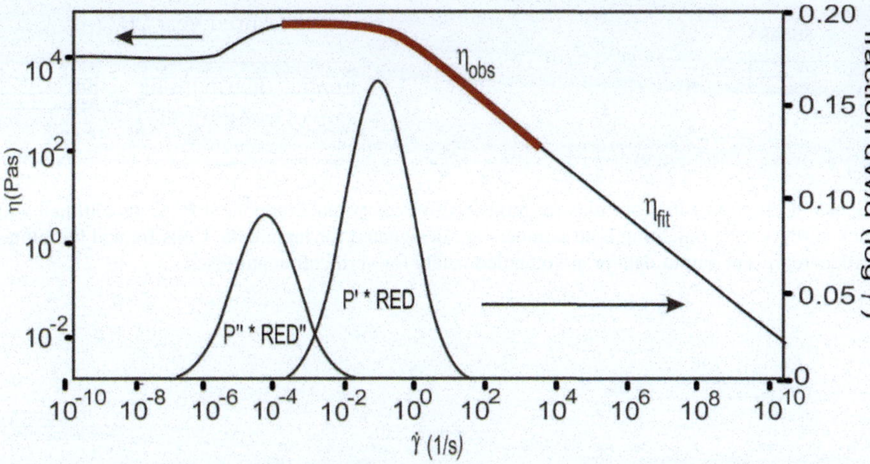

Fig. 4.7 On the MS Excel 365 spreadsheet cut and redrawn chart as it is simulation for the same polymer as in the Fig. 4.4 but used instead Wesslau distribution with the sets $\mu = -1.4$ and $\sigma = 0.9$. Also $P' = 0.40$, $P'' = 0.22$ and $\log R = 3.2$ are the same, which closest fits the shear viscosity measurements by Meissner and Laun, as shown in the thick segment of the curve

4.14.1 Spreadsheet Data for Modelling Viscosity

Attached is cut of pre-prepared calculation spreadsheet with formulas of the same computation as get Fig. 4.7. In the Sect. 3.10 is explained some general principles for presented spreadsheets. Figures 4.8 and 4.9 show same cells in the Formula Auditing Mode (see Fig. 4.10).

	B	C	D	E	F	G	H	I	J	K	L	M	N	O	P	Q	R	S
3		SHEAR VISCOSITY																
4																		
5	μ	**-1.4**		R'		0	P'	**0.4**		%RMSE	0.34	%						
6	σ	**0.9**		R''		3.2	P''	**0.22**										
7							log eta	**4.05**										0.34
8	x=Shear Rate (gammaDOT)								1.00					1.00			1840H	0.06
9	row	x	log x	w	log x'	tilde w(log x)	RED'	int (log x')	P' (log x	log x''	tilde w'	RED''	int (log x'')	P'' (log x''	P(x)	η/fit	η/obs	
10	0	0.00	-10	6.5955E-21	-10	6.5955E-21	3E-21	0.00	0.00	-6.80	0.00	3E-09	0	0.00	5.49E-10	4.05		
11	1	0.00	-9.9	1.8953E-20	-9.9	1.8953E-20	8E-21	0.00	0.00	-6.70	0.00	5E-09	0.00	0.00	5.49E-10	4.05		

Fig. 4.8 Presents a top section of an MS Excel 365 spreadsheet used for modelling viscosity. Cells outlined with thick borders and containing bold numbers are designated for input data. The RheoPower software internally calculates these values when processing viscosity measurement data. Cells framed by thinner borders represent output data or are intended solely for verification purposes. The Wesslau distribution, calculated by formula, is already normalized. Additionally, RED is normalized if the procedure presented in Sect. 7.9, "Importing MWD Data on a Spreadsheet," is utilized

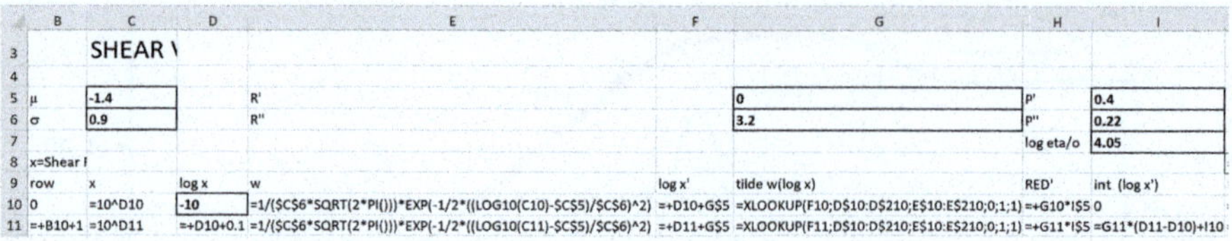

	B	C	D	E	F	G	H	I
3		SHEAR \						
4								
5	μ	-1.4		R'		0	P'	0.4
6	σ	0.9		R''		3.2	P''	0.22
7							log eta/o	4.05
8	x=Shear I							
9	row	x	log x	w	log x'	tilde w(log x)	RED'	int (log x')
10	0	=10^D10	-10	=1/($C$6*SQRT(2*PI()))*EXP(-1/2*((LOG10(C10)-$C$5)/$C$6)^2)	=+D10+G$5	=XLOOKUP(F10;D$10:D$210;E$10:E$210;0;1;1)	=+G10*I$5	0
11	=+B10+1	=10^D11	=+D10+0.1	=1/($C$6*SQRT(2*PI()))*EXP(-1/2*((LOG10(C11)-$C$5)/$C$6)^2)	=+D11+G$5	=XLOOKUP(F11;D$10:D$210;E$10:E$210;0;1;1)	=+G11*I$5	=G11*(D11-D10)+I10

Fig. 4.9 Left part of the formulas for modelling viscosity and component *P′*. Cells outlined with thick borders and containing bold numbers are designated for input data. Cells framed by thinner borders represent output data or are intended solely for verification purposes

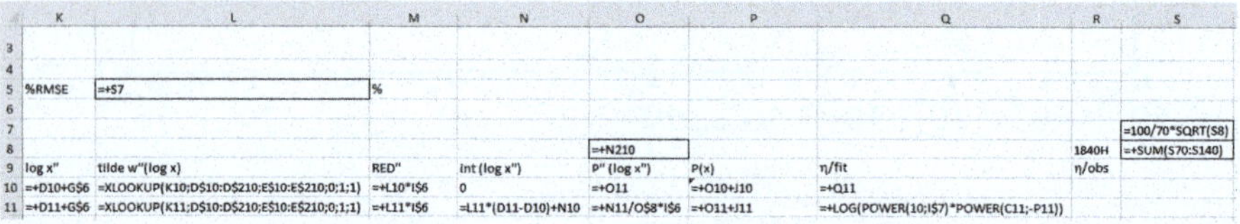

	K	L	M	N	O	P	Q	R	S
3									
4									
5	%RMSE	=+S7	%						
6									
7									=100/70*SQRT(S8)
8					=+N210			1840H	=+SUM(S70:S140)
9	log x''	tilde w''(log x)	RED''	int (log x'')	P'' (log x'')	P(x)	η/fit	η/obs	
10	=+D10+G$6	=XLOOKUP(K10;D$10:D$210;E$10:E$210;0;1;1)	=+L10*I$6	0	=+O11	=+O10+J10	=+Q11		
11	=+D11+G$6	=XLOOKUP(K11;D$10:D$210;E$10:E$210;0;1;1)	=+L11*I$6	=L11*(D11-D10)+N10	=+N11/O$8*I$6	=+O11+J11	=+LOG(POWER(10;I$7)*POWER(C11;-P11))		

Fig. 4.10 Right side part of the formulas for modelling and component *P″* and complete viscosity. Columns T and U with row 11 can be demonstrating data and formulas for comparing measured and observed with modelled viscosity

4.14.2 Formulas of the Elastic Component on the Spreadsheet

Input values:

C5 −1.4 (μ is the mean)
F5 0 (*R′* the place of RED′ at a rate scale for copy of RED)
I5 0.40 (*P′* elastic value)
C6 0.9 (σ is standard deviation)
F6 3.2 (log *R*, *R″* the place of RED″ at a rate scale for copy of RED)
I6 0.22 (*P″* viscosity value)
I7 4.05 (Zero viscosity η_0 at shear rate 1/s)
D10 −10 (initial log rate for rows and arrays).

Formulas:

J8 =+I210
B11 =+B10 + 1
C11 =10^D11
D11 =+D10 + 0.1
E11 =1/($C$6*SQRT(2*PI()))*EXP(−1/2*((LOG10(C11) − $C$5)/$C$6)^2)
F11 =+D11 + G$5
I11 =G11*(D11 − D10) + I10
J11 =+I10/J$8*I$5.

4.14.3 Formulas of the Viscous Component on the Spreadsheet

Formulas:

O8 =+N210
K11 =+D11 + G$6
L11 =XLOOKUP(K11;D$10:D$210;E$10:E$210;0;1;1)
M11 =+L11*I$6
N11 =L11*(D11 – D10) + N10
O11 =+N11/O$8*I$6
P11 =+O11 + J11
Q11 =+LOG(POWER(10;I$7)*POWER(C11; –P11)).

Option input and formulas:

L5 =+S7
S7 =100/(η_{OBS} items number)*SQRT(U8)
S8 =+SUM(Sa:Sb) first row Sa and Sb of η_{OBS} data for %RMSE error
S100 =((10^Q100 – 10^R100)/(10^Q100))^2.

4.15 Conclusions

We can model viscosity flow curves accurately at error level of %RMSE = 0.27 by using Wesslau distributions with the form of Normal (Gaussian) on logarithmic scale standard monomodal polymers.

By using modified MWD form we can achieve ten times less error in the range of %RMSE = 0.03.

But if we compute which can modify MWD during computations related to measurements, the error is as less as below %RMSE < 0.003 as discussed in the Sect. 13.8.

Formulas of the Analytical Model, which are directly related to Control Theory, are presented initially in all chapters to provide a background and platform for the Unified Model. This model is based on the discovery that not all molecules respond to changes in shear rate. Although the Unified Model yields more accurate results and is easier to use, it may be found in the future that some molecules are slightly active outside the boundaries of the Unified Model.

References

1. Borg, T., & Pääkkönen, E. J. (2009). Linear viscoelastic models. Part II. Recovery of the molecular weight distribution using viscosity data. *Journal of Non-Newtonian Fluid Mechanics, 156*, 129–138.
2. Meissner, J. (1975). Basic parameters, melt rheology, processing and end-use properties of three similar low-density polyethylene samples. *Pure and Applied Chemistry, 42*, 551–662.
3. Laun, H. M. (1982). Elastic properties of polyethylene melts at high shear rates with respect to extrusion. *Rheologica Acta, 21*, 464–469.
4. Borg, T., & Pääkkönen, E. J. (2009). Linear viscoelastic models. Part I. Relaxation modulus and melt calibration. *Journal of Non-Newtonian Fluid Mechanics, 156*, 121–128.
5. Graessley, W. W. (2008). *Polymetric liquids and networks: Dynamics and rheology*. Garland Science.
6. Anderssen, R. S., & Loy, R. J. (2001). On the scaling of molecular weight distribution functionals. *Journal of Rheology, 45*, 891–901.
7. Rouse, P. R., Jr. (1953). A theory of the linear viscoelastic properties of dilute solutions of coiling polymers. *The Journal of Chemical Physics, 21*, 1272–1280.
8. Bruckner, G., & Cheng, J. (2000). Tikhonov regularization for an integral equation of the first kind with logarithmic kernel. *J. Inverse and Ill-Posed Problems, 8*, 665–675.
9. Arfken, G., & Weber, H. (2001). *Mathematical methods for physicists* (5th ed.). Academic.
10. Poh, L., Wu, Q., Chen, Y., & Narimissa, E. (2022). Characterization of industrial low-density polyethylene: A thermal, dynamic mechanical, and rheological investigation. *Rheologica Acta, 61*, 701–720.

Chapter 5
Complex Viscosity

We have developed a steady shear viscosity formula based on Control Theory and *Unified* principles. Now, we are extending this methodology to complex viscosity and modulus [1].

The *Unified Model* was referred to as the Characteristic Model in our published paper. The complex modulus G^* is as the instantaneous ratio of the stress to strain and is defined in Eq. 5.1.

$$G^*(\omega) = G'(\omega) + iG''(\omega) \tag{5.1}$$

Complex viscosity $\eta^*(\omega)$ is defined according to Eq. 5.2.

$$\left|\eta^*(\omega)\right| = \left|\frac{G^*(\omega)}{\omega}\right| = \frac{1}{\omega}\sqrt{G'^2(\omega) + G''^2(\omega)} \tag{5.2}$$

In this section, we model $\eta^*(\omega)$ to correspond with the discussions in Chapter 4. The presentation of practical formulas begins with Eq. 5.5.

5.1 CT Principle for Complex Shear Viscosity

The response complex viscosity η^* (ω) is derived from response $h(\psi)$ by sampling the active molecules in the distribution $w(\psi)$ over some shear rate interval, as follows:

$$\eta^*(\omega) = \int_{-\infty}^{\omega} w(\psi)h(\omega - \psi)d\psi \tag{5.3}$$

T. Borg, *Control Theory in Rheology*, https://doi.org/10.1007/978-3-031-88248-7_5

Here, ψ is the standard integration variable for frequency rate ω. This equation is a linear formula used in control theory to adjust and control a single variable or function in a closed system. We applied this formula to determine the $w(\omega)$ distribution function and molecular weight distribution (MWD), as shown later in the following sections. Molar background for principle is discussed in the Chap. 6.

As the molecular weight distribution (MWD), denoted as $w(\log M)$, is typically represented in fractions on a logarithmic scale, or here RED $w(\log \omega)$, we have to rewrite all variables and functions in Eq. 5.3 on a logarithmic scale:

$\log \omega - \log \psi = \log \frac{\omega}{\psi}$, and $d\psi$ is written as $d \log \psi$.

The logarithmic convolution (also known as the scale convolution) can be related to ordinary convolution by taking the logarithm of ω. We can, therefore, write $h(\omega) = \omega$ and d ω as d(log ω). Response viscosity $\eta^*(\omega)$ is a normalized viscosity $log\frac{\eta^*(\omega)}{\eta_0^*}$ on the logarithmic scale at ω_0. The zero viscosity at frequency 1/s, η_0^*, is obtained by fitting to experimental measurements. We then obtain the general complex shear viscosity function $\eta(\omega)$:

$$\log \frac{\eta^*(\omega)}{\eta_0^*} = -P' \int_{-\infty}^{\log \omega} w(\log \psi) \log \frac{\omega}{\psi} d \log \psi \tag{5.4}$$

When we include also the viscous Rouse shear, which is related to relaxation with orientation or entanglement shear, the complete viscoelastic formula for $\eta^*(\omega)$ at different frequencies can be expressed common logarithm distributions as

$$\log \frac{\eta^*(\omega)}{\eta_0^*} = - \int_{\log \omega/T}^{\log \omega} \left(P'w'(\log \psi) + P''w''\left(\log \frac{\psi}{R}\right)\right) \log \frac{\omega}{\psi} d \log \psi \tag{5.5}$$

where the elastic value P' is now in the entanglement or orientation range known also as plateau region and viscous P'' is in the Rouse shear range known also as terminal zone. This is achieved simply by copying function $w(\log \omega)$ after dividing ω by the ratio, R, of the frequency-rate ranges to obtain the Rouse distribution, $w(\log\frac{\omega}{\psi})$. During steady state frequency the greatest molecules are no more effective for the viscosity and we have to use the lower bound of integration $\log \omega - \log T = \log\frac{\omega}{T}$ instead of the minus infinity limit. The sampling band is still wide in the range $T =$ 100,000/s. Another reason for introducing the sampling band T is that, in oscillating rheometers, the actual shear rates experienced by the molecules vary significantly, and the frequency ω serves as a virtual integral of all shear rates at the ω frequency.

The model described by Eq. (5.3) includes complex viscosity $\eta^*(\omega)$ without difficulty, but solving distribution $w(\omega)$ is known to be a severely ill-posed problem reported by Bruckner and Cheng [2] due to used kernel $\log\frac{\omega}{\psi}$ inside convolution integral equation as discussed in Chap. 4 and 6.

To maintain consistency among constants and variables, we need to normalize $\tilde{w}(\log \omega)$ using the following principle:

General normalization of the cumulative distribution for RED $\tilde{w}(\log \omega)$ can be written in shortly form shown Eq. 5.6 as follows:

$$\int_{-\infty}^{\infty} w(\log \omega) d \log \omega = 1 \tag{5.6}$$

We present normalization procedure in the Sect. 3.8 also for spreadsheet calculations.

5.2 Unified Model

We use a similar procedure as for shear viscosity Eq. 4.7 to generate a Unified formula for complex viscosity $\eta^*(\omega)$ as a function of frequency ω:

$$\log \frac{\eta^*(\omega)}{\eta_0^*} = -\log \frac{\omega}{\omega_c} \int_{-\infty}^{\log \omega} \left(P' w'(\log \psi) + P'' w'' \left(\log \frac{\psi}{R} \right) \right) d \log \psi \tag{5.7}$$

Here, response $\log \omega - \log \psi = \log \frac{\omega}{\psi}$, and ψ is the standard integration variable. While the characteristic frequency rate ω_c is typically 1/s in most cases, it may vary slightly around 1/s during MWD computation for practical reasons explained later. Finally, we obtain the equation for $\eta^*(\omega)$:

$$\eta^*(\omega) = \eta_0^* 10^{-\log \frac{\omega}{\omega_c} \int_{-\infty}^{og\omega} \left(P' w'(\log \psi) + P'' w'' \left(\log \frac{\psi}{R} \right) \right) d \log \psi} \tag{5.8}$$

We use the common logarithm for the left side of Eq. (5.7) $\log \frac{\eta^*(\omega)}{\eta_0^*}$, to derive the power law formula directly. Distribution integrals are presented in common logarithmic form as well, facilitating the determination of polymer structure constants for elastic P' and viscous P'' that are comparable to each other.

In the Fig. 5.1 shown P' and P'' values are found originally by RheoPower software during thousands of iterations for data measured at 150 °C. At first is derived the measured complex viscosity data range shown thick red colour to get piece of respective RED segment. Then is added the first guess to the RED sides to get compete distribution. Now is computed fit viscosity curve and compared to the measured viscosity. On the next computation loop is changed a little the both sides of RED distribution and is made new viscosity curve model. After thousands of loops is found the lowest error between measured and modelled viscosity curve as described procedure in the Chapter 9. Now the obtained RED is converted to the MWD by melt calibration factors Mf and Hf, which are constant to the similar polymer types at a temperature. %RMSE = 0.48%.

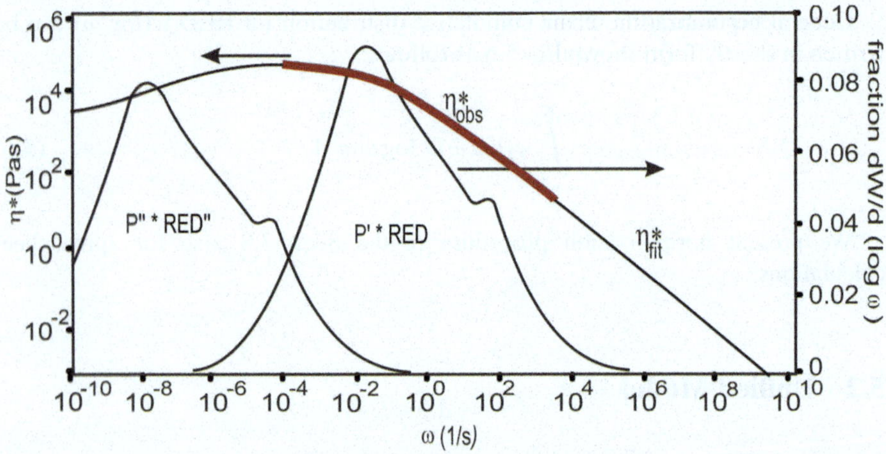

Fig. 5.1 Complex viscosity $\eta^*(\omega)$ at 150 °C temperature modelled by 1840H LDPE. Original data is measured by Basf and RheoPower computed RED, which the base form of distribution is used for all simulations with 1840H LDPE. MWD data is based year 2005 information, which gave for MWD $Mf = 26{,}700$ g/mol and $Hf = 2.72$ to the relation to RED

5.3 Cumulative Polymer Structure Function

For simplicity, we write Eq. 5.8 as an exponent without relation originally kernel of convolution integral $-\log\frac{\omega}{\omega_c}$ and set up a very practical cumulative polymer structure function $P(\log\omega)$:

$$P(\log\omega) = \int_{-\infty}^{\log\omega} \left(P'w'(\log\psi) + P''w''\left(\log\frac{\psi}{R}\right)\right) d\log\psi \tag{5.9}$$

where cumulative integral components are:

$$P'(\omega) = P' \int_{-\infty}^{\log\omega} w'(\log\psi)d\log\psi \tag{5.10}$$

$$P''(\omega) = P'' \int_{-\infty}^{\log\omega} w''\left(\log\frac{\psi}{R}\right)d\log\psi \tag{5.11}$$

Although absolute value level of rheologically effective distributions (REDs) $w'(\log\omega)$ and $w''(\log\omega)$ have no effect thanks to normalization of cumulative integral, a good habit is to use normalized before use in the CT principle. RED can be in the form $w(\log\dot{\gamma})$, $w(\log t)$ or $w(\log\dot{\varepsilon})$. But respective integral functions $P'(\log\omega)$ and $P''(\log\omega)$ and as functions of $\dot{\gamma}$, t and $\dot{\varepsilon}$ must be scaled similar way before multiplying by constants P' and P''.

According to the used principle, $P(\log\omega)$ indicates the influence of effective viscous $w''(\log\omega)$ and also changes in the elastic $w'(\log\omega)$ distribution REDs.

Besides incorporating the chain structure, it relates the orientation and degree of entanglement as functions of the shear rate and respective frequency (Figs. 5.2 and 5.3).

It is much more convenient to use the short form of the equations. We write polymer structure function $P(\omega)$ with non-logarithmic variable as follows:

$$P(\omega) = P'(\omega) + P'(\omega) \tag{5.12}$$

This gives us a simple power-form formula to Eq. 5.8 included relation $-\log\frac{\omega}{\omega_c}$ with $P(\omega)$ to get complete original formula and approximates the true molecular

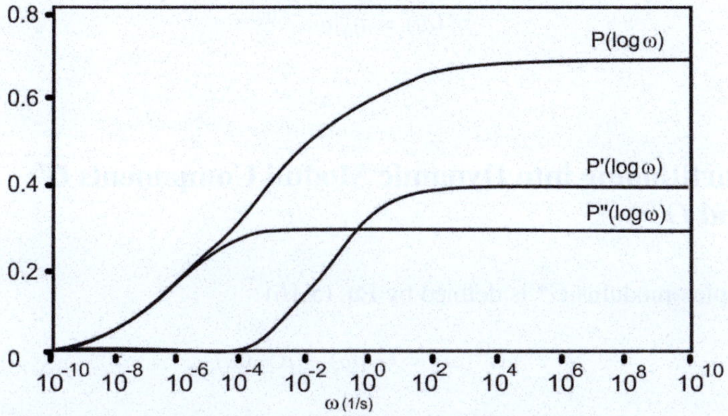

Fig. 5.2 Cumulative elastic $P'(\log \omega)$, viscous $P''(\log \omega)$ integral components and cumulative polymer structure function $P(\log \omega)$ for LDPE H1840. Practical model is used by Unified Eq. 5.8

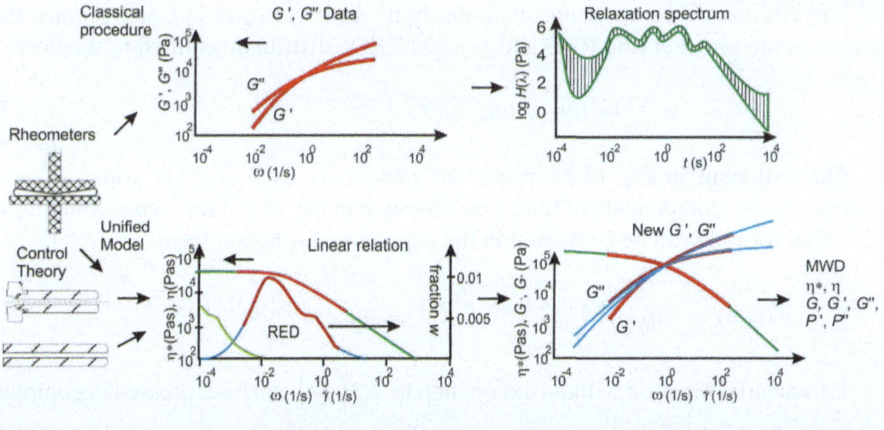

Fig. 5.3 Schematic comparison of the classical and new procedures [3]. Thicker and thinner lines indicate measured and modelled data, respectively. Note that modelled *G'* has slightly higher values and *G''* has slightly lower values at low frequencies even though *G** is based on data measured

orientation level and state in Eq. (5.13):

$$\eta^*(\omega) = \eta_0^* 10^{-\log\frac{\omega}{\omega_c}P(\omega)} = \eta_0^*\left(10^{\log\frac{\omega}{\omega_c}}\right)^{-P(\omega)} \tag{5.13}$$

As we used common logarithm, we can write directly Eq. (5.14).

$$\eta^*(\omega) = \eta_0^*\left(\frac{\omega}{\omega_c}\right)^{-P(\omega)} \tag{5.14}$$

Finally, similar way as explained with (Eq. 5.15) we can write power law formula:

$$\eta^*(\omega) = \eta_0^*|\omega|^{-P(\omega)} \tag{5.15}$$

5.4 Partitioning into Dynamic Moduli Components G' and G''

The complex modulus G^* is defined by Eq. (5.16)

$$G^*(\omega) = G'(\omega) + iG''(\omega) \tag{5.16}$$

where G′ is the elastic component and G" is the viscous component. As Control Theory (CT) and Unified models utilize molecular orientation and shear viscosity as a foundation, we compute G′ and G″ starting from the complex viscosity formula. We model the complex modulus $G^* = \eta^*\omega$. Additionally, we partition into elastic G' and viscous G'' components analogously to shear viscosity in Chap. 4 since the source is the same: elastic RED and viscous RED″ distributions fitted to the flow.

$$G^*(\omega) = \omega\eta_0^* 10^{\log\frac{\omega}{\omega_c}(-P'(\omega)-P''(\omega))} \tag{5.17}$$

The exponent in Eq. (5.8), presented previously as $-\log\frac{\omega}{\omega_c}\left(P'(\omega)+P''(\omega)\right)$ comprises two components of functions rewritten in Eq. (5.17) for complex modulus G^*. This formula can be converted in the exponential function form:

$$G^*(\omega) = \omega\eta_0^* 10^{\log\frac{\omega}{\omega_c}(-P'(\omega)-P''(\omega))} = \omega\eta_0^* e^{2.3\log\left(\frac{\omega}{\omega_c}\right)(-P'(\omega)-P''(\omega))}.$$

Exponential function without exponent part $2.3\left(\frac{\omega}{\omega_c}\right)$ can be expressed in complex notation as $e^{-P'(\omega)-P''(\omega)}$, where functions $P'(\omega)$ and $P''(\omega)$ are given in Eqs. (5.10) and (5.11).

The obtained formula bears similarities to complex notation. Therefore, we define the exponential of the complex variable z as $e^z = e^{-P'(\omega)-iP''(\omega)}$, a concept discussed earlier in the context of shear viscosity in Sect. 5. 4 and applied to oscillatory shear models.

We then partition Eq. (5.17) into components, where the part of exponent $2.3\log\left(\frac{\dot{\gamma}}{\dot{\gamma}_c}\right)$ is divided between the numerator and denominator. The elastic *G'* and viscous *G"* are defined as:

$$G'(\omega) = G^*(\omega)\frac{P'(\omega)}{\sqrt{P'^2(\omega) + P''^2(\omega)}} \tag{5.18}$$

$$G''(\omega) = G^*(\omega)\frac{P''(\omega)}{\sqrt{P'^2(\omega) + P''^2(\omega)}} \tag{5.19}$$

A complex number is nothing more than an ordered pair of two real numbers [4] and functions are also real-valued. Similar partitioning is applied to the shear viscosity in the earlier Chap. 4.

This partitioning can also be applied to the shear viscosity and relaxation modulus or used with the computed dynamic storage *G'* and loss *G"*.

moduli as measured data for the familiar deconvolution procedure to obtain $G(t)$. However, difficulties arise when modelling certain polyolefin measurements at low temperatures due to long transition time effects. Borg and Pääkkönen [5].

5.5 Practical Applications

The viscosity of Lupolen 1840H LDPE measured at temperatures ranging from 130 to 250 °C is shown in Fig. 5.4. The points represent measured data provided by BASF, and the wide-range modelled viscosity curves are based on computations as described in Chap. 9. The main goal of several iterations was to detect and get MWD of polymer.

Lupolen 1840H LDPE by BASF was selected as the LDPE sample since measured data were available and the material is widely reported in the literature. The sample was a modern version of the classical LDPE IUPAC A, or "Melt I," first reported by Meissner [6, 7]. Measured MWD by GPC/SEC has changed during years as discussed in Sect. 3.8, but may have different MWDs during years, the RED is always constant related to the viscoelastic properties.

The MWDs of LDPE computed at 13 temperatures are summarized in Figs. 5.5 and 5.4. This test revealed that temperature had a minor effect on the values of the constants:

$$Mf = Mf_0 + (T - T_0)250 \tag{5.20}$$

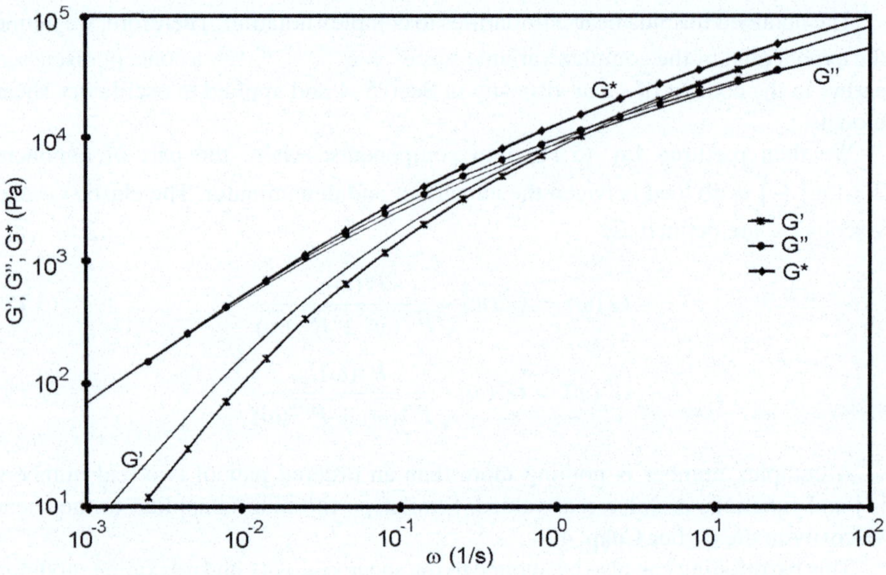

Fig. 5.4 Measured data (shown by symbols) and modelled storage *G'*, loss *G''* and complex *G** moduli for 1840H LDPE (a modern version of IUPAC A) at 150 °C. Similar results were obtained at higher temperatures, but the discrepancy between *G'* and *G''* increased at lower temperatures. Additionally, at higher frequencies, the measured and modelled values of *G'* and *G''* are consistent, as shown in the figure

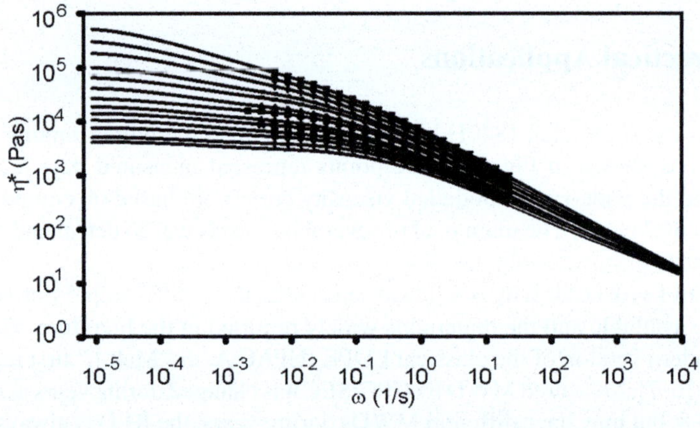

Fig. 5.5 Measured viscosity curves for Lupolen 1840H LDPE at temperatures ranging from 130 °C to 250 °C in 10 °C increments. The points represent measured data provided by BASF, and the curves shown over a wider scale were modelled using software. The average error (%RMSE) between the measured and modelled viscosities is 0.0009%. Borg and Pääkkönen [1]

Table 5.1 Main characteristics of all investigated samples and computations

	Ta	Mw[b]	MwR[b]	Mf[c]	Hf[c]	P'[d]	P''[d]	%RMSE[e]
1840H	130	214 404	13.3	32 508	2.05	0.21	0.43	0.0011
1840H	140	235 996	14.3	35 532	2.05	0.28	0.38	0.0007
1840H	150	239 973	14.6	37 800	2.05	0.26	0.39	0.0011
1840H	160	238 833	14.2	40 068	2.05	0.28	0.35	0.0007
1840H	170	239 191	14.3	42 092	2.05	0.32	0.32	0.0007
1840H	180	239 505	14.1	45 360	2.05	0.35	0.29	0.0007
1840H	190	239 580	14.4	47 628	2.05	0.35	0.27	0.0007
1840H	200	237 497	14.0	50 652	2.05	0.37	0.25	0.0007
1840H	210	240 150	14.2	52 920	2.05	0.38	0.25	0.0007
1840H	220	215 824	13.5	55 188	2.05	0.41	0.21	0.0011
1840H	230	236 195	14.4	57 456	2.05	0.34	0.24	0.0012
1840H	240	239 123	14.3	60 748	2.05	0.45	0.17	0.0011
1840H	250	203 579	12.5	62 748	2.05	0.41	0.16	0.0015

[a] Measured temperature in Celsius °C
[b] Computed average molecular weight *Mw* in g/mol and polydispersity index $Mw/Mn = MwR$
[c] Used structural value *Mf* and conversion factor *Hf*. Information not always available, if the used rheometer type was plate-plate or cone-plate
[d] Obtained or used elasticity and viscosity values P' and P''
[e] Obtained percentage root-mean-square error function %RMSE

The main characteristics of these samples are listed in Table 5.1. Please obtain that all other values are generated by software during interactions between modelled viscosity and computed MWD as elastic P' and viscous P'' values and MWD with *Mw* and *MwR* results. The average error (%RMSE) between measurements and modelled viscosity is obtained by Eq. 9.7.

5.6 Converting by Melt Calibration Equations Between Scales

Converting the normalized rheologically effective distribution RED $w(\log\,\omega)$ to molecular weight distribution MWD $\tilde{w}(\log M)$ is discussed in the Melt Calibration Chapter 6, but we already present formulas. For shear viscosity ω, conversion equations are as follows to get from MWD to RED:

$$w(\log M) = \tilde{w}(\log \omega)\Big|_{\omega=\omega_1\left(\frac{Mf}{M}\right)^{Hf}} \tag{5.21}$$

To get from RED to MWD:

$$w(\log\omega) = \tilde{w}(\log M)\Big|_{M=Mf\left(\frac{\omega}{\omega_1}\right)^{-\frac{1}{Hf}}} \quad (5.22)$$

A good practise is to normalize $\tilde{w}(\log\omega)$ function as explained in Sect. 3.8 before proceeding further.

5.7 Cox-Merz Rules

The original Cox-Merz rule [8] $\eta(\dot{\gamma}) = |\eta^*(\omega)|_{\omega=\dot{\gamma}}$ and its extended or modified variations can be explained using the presented principle. Different flow scales of t, ω, and $\dot{\gamma}$ do not have direct physical interrelations. Sometimes Cox-Merz rules hold, and sometimes they fail completely, as investigated by many studies. It is not always possible to find the correct MWD from control theory, but very rarely have the viscoelastic properties not been modelled accurately. For these different flow types, the same polymer and corresponding MWD are used, but the flow deformations differ.

We can use the MWD obtained from melt calibration to generate Rheological Equivalence Distribution (REDs) as a function of shear rate, $\dot{\gamma}$, or $w(\log M) = w'$ $(\log\dot{\gamma})$ or as a function t and ω, using different P' and P'' constants and effective zero states $\hat{\eta}_0$, $\hat{\eta}_0^*$ or $\hat{G}_0$ depending on the flow type. Thus, we can convert viscoelastic data between different deformation modes, explaining Cox-Merz rules when P', P'', and the zero states are the same:

$$w(M) = |\eta(\dot{\gamma})|_{\eta_0,P',P''};\ |\eta^*(\omega)|_{\eta_0^*,P',P''};\ |G(t)|_{G_0,P',P''} \quad (5.23)$$

In other words, the same constant (MWD $w(M)$) is used as the starting and reference point for different REDs and flow scales of t, ω, and $\dot{\gamma}$, as shown in Fig. 5.5. When P', P'', and the zero states are the same, Cox-Merz rules hold. The use of frequency ω (1/s) or angular frequency ω (rad/s) could also be adjusted by the structural factor Mf to obtain the correct fit, with a similar approach being taken for modified or extended Cox-Merz rules.

Our presented principle based on melt calibration is much more versatile and allows the conversion to any linear scales by adjusting Mf and scale units using the conversion factor Hf.

We used here Lupolen 1840H LDPE by BASF and its ancestor "Melt I" in presentation as they are lot of studied and different type data is available. This is also example where Cox-Merz rule holds well between shear and complex viscosity. The relation is not used anywhere in this book.

In summary, the primary reason for Cox-Merz rules holding sometimes is that having the same MWD provides similarities in the form of curves, but in other cases, it is a matter of chance. Nevertheless, the Unified and CT principles demonstrates that it is occasionally possible.

5.8 Examples

To approximate the wide function for shear viscosity, $\eta^*(\omega)$, we use a normalized Wesslau distribution for $w(\log \omega)$ for RED. We then fit the appropriate form and position of the distribution using constants η_0^*, P', and P'' in the Sect. 3.11 "Spreadsheet Examples.“ Further discussion on the use of spreadsheets can be found in Sect. 4.13.

Complex viscosity $\eta^*(\omega)$ and modulus $G^*(\omega)$ are used with oscillating rheometers and flows. Conversion for complex viscosity $\eta^*(\omega) = \frac{G^*(\omega)}{2\pi\omega}$, if complex modulus is measured in angular frequency [rad/s] scale.

Databases on the RheoPower software are shear rate [1/s], frequency [Hz] and time [s] modes.

5.8.1 *On a Spreadsheet*

We model the true complex viscosity measurements of 1840H LDPE using artificial Wesslau distribution for REDs, as shown in Fig. 5.6, with closely matching average molecular weight and polydispersity values to those of 1840H LDPE. For Wesslau distribution is found the best fit viscosity by values $\mu = -0.9$ and $\sigma = 1.92$ to those modelled from measurements by Basf. The elastic value $P' = 0.39$ and viscous $P'' = 0.31$. No doubt, measured MWD for 1840H LDPE can be utilized, but this data sample would require 200 rows for display in subsequent spreadsheets.

5.8.2 *Formulas on the Spreadsheet*

Attached Fig. 5.8 is cut of pre-prepared calculation spreadsheet with 7formulas of the same computation as get Fig. 5.7. In the Sect. 3.10 is explained some general principles for presented spreadsheets.Figures 5.9 and 5.10 show same cells in the Formula Auditing Mode.

Input values:

C5 –0.9 (μ is the mean).
G5 0 (*R'* the place of RED*'* at a rate scale for copy of RED).
I5 0.39 (*P'* elastic value).
C6 1.92 (σ is standard deviation).
G6 6 (log *R*, *R''* the place of RED*''* at a rate scale for copy of RED).
I6 0.31 (*P''* viscosity value).
I7 3.684 (Zero viscosity η^*_0 at frequency rate 1/s).
D10 –10 (initial log rate for rows and arrays).

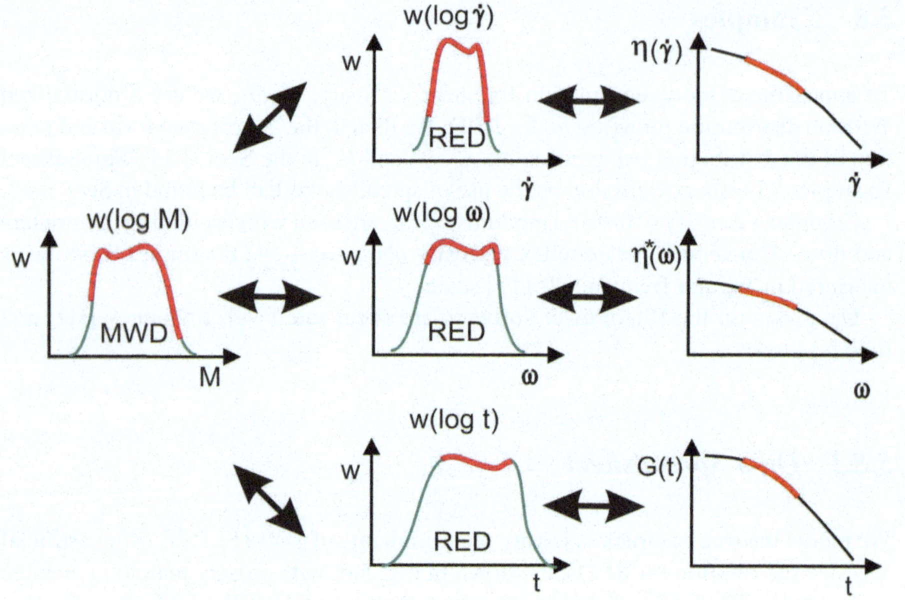

Fig. 5.6 Using the presented method of melt calibration yields the MWD that remains the same for different flows and scales. Thicker lines represent measured data. The main point is the fact that shear viscosity measurements can scan much wider RED/MWD as other traditional procedures. Not so important information is the fact that the original Cox-Merz rule holds if P', P'', RED and the zero states are the same in different flows. Borg and Pääkkönen [3]

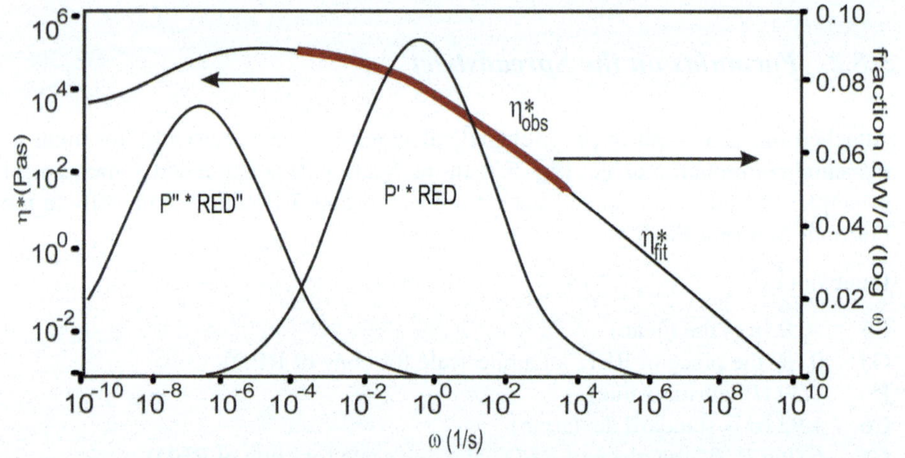

Fig. 5.7 On the MS Excel 365 spreadsheet cut and redrawn chart as it is simulation for the same polymer as in the Fig. 5.1 but used instead Wesslau distribution with the sets μ = -0.9 and σ = 1.92. $P' = 0.39$, $P'' = 0.31$, $\log R = 6$ are the same with Fig. 5.1, which closest fits the complex viscosity measurements by Basf at 150 °C

	B	C	D	E	F	G	H	I	J	K	L	M	N	O	P	Q	R	S
3		Complex Viscosity																
4																		
5	μ	-0.9			R'	0	P'	0.39		%RMSE	0.59	%						
6	σ	1.92			R''	6	P''	0.31										
7							log eta*	3.648										
8		x= frequency (Hz)							1.00					0.94				0.17
9	row	x	log x	w	log x'	tilde w(log x)	RED' w(log x)	int (log x')	P' (log x')	log x''	tilde w''(	RED''	int (log x'')	P'' (log x'')	P(x)	log eta*	Measured	0.59
10	0	0.00	-10	2.8E-06	-10	2.75226E-06	1.0734E-06	0.00	0.00	-4.00	0.06	0.0174942	0	0.00	2.01E-03	3.67		
11	1	0.00	-9.9	3.5E-06	-9.9	3.51809E-06	1.3721E-06	0.00	0.00	-3.90	0.06	0.0190032	0.01	0.00	2.02E-03	3.67		

Fig. 5.8 Presents a top section of an MS Excel 365 spreadsheet used for modelling complex viscosity. Cells outlined with thick borders and containing bold numbers are designated for input data. The RheoPower software internally determines these values when computing viscosity measurements data. Cells framed by thinner borders represent output data or are intended solely for verification purposes. The Wesslau distribution, calculated by formula, is already normalized. Additionally, RED is normalized if the procedure presented in Sect. 7.9, "Importing MWD Data on a Spreadsheet," is utilized

	B	C	D	E	F	G	H	I	J
1									
2									
3		Comp							
4									
5	μ	-0.9			R'	0	P'	0.39	
6	σ	1.92			R''	6	P''	0.31	
7							log eta*	3.648	
8		x= freque							=+I210
9	row	x	log x	w	log x'	tilde w(log x)	RED' w(log x)	int (log x')	P' (log x')
10	0	=10^D10	-10	=1/($C$6*SQRT(2*PI()))*EXP(-1/2*((LOG10(C10)-$C$5)/$C$6)^2)	=+D10+G$5	=XLOOKUP(F10;D$10:D$210;E$10:E$210;0;1;1	=+G10*I$5	=+G10	=+I10/J$8*I$5
11	=+B10+1	=10^D11	=+D10+0.1	=1/($C$6*SQRT(2*PI()))*EXP(-1/2*((LOG10(C11)-$C$5)/$C$6)^2)	=+D11+G$5	=XLOOKUP(F11;D$10:D$210;E$10:E$210;0;1;1	=+G11*I$5	=G11*(D11-D10)+I10	=+I11/J$8*I$5

Fig. 5.9 Left part of the formulas for modelling viscosity and component P'. Cells outlined with thick borders and containing bold numbers are designated for input data. Cells framed by thinner borders represent output data or are intended solely for verification purposes

	K	L	M	N	O	P	Q	R	S	T	U
3											
4											
5	%RMSE	=+S9	%								
6											
7											
8					=+N210				=+SUM(S90:S120)		
9	log x''	tilde w''(log x)	RED''	int (log x'')	P'' (log x'')	P(x)	log eta*	Measured	=100/70*SQRT(S8	log x	Measured
10	=+D10+G$6	=XLOOKUP(K10;D$10:D$210;E$10:E$210;0;1;1)	=+L10*I$6	0	=+O11	=+O10+J10	=+Q11			-4	4.96370756986293
11	=+D11+G$6	=XLOOKUP(K11;D$10:D$210;E$10:E$210;0;1;1)	=+L11*I$6	=L11*(D11-D10)+N10	=+N11/O$8*I$6	=+O11+J11	=+LOG(POWER(10;I$7)*POWER(C11;-P11))			-3.95000714307986	4.95309252059755

Fig. 5.10 Right side part of the formulas for modelling and component P'' and complete complex viscosity. Columns R and S with row 11 is demonstrating data and formulas for comparing measured and observed with modelled viscosity

Formulas:

J8 = + I210.
B11 = + B10 + 1.
C11 = 10^D11.
D11 = + D10 + 0.1
E11 = 1/($C$6*SQRT(2*PI()))*EXP(-1/2*((LOG10(C11)-$C$5)/$C$6)^2).
F11 = + D11 + G$5.
G11 = XLOOKUP(F11;D$10:D$210;E$10:E$210;0;1;1).
H11 = + G11*I$5.
I11 = G11*(D11-D10) + I10.
J11 = + I11/J$8*I$5.

Formulas:

O8 = + N210.
L11 = + D11 + G$6.
M11 = + L11*I$6.
N11 = L11*(D11-D10) + N10.
O11 = + N11/O$8*I$6.
P11 = + O11 + J11.
Q11 = + LOG(POWER(10;I$7)*POWER(C11;-P11)).

Option input and formulas:

L5 = + S9.
S8 = + SUM(Sa:Sb) first row Sa and Sb of η_{OBS} data for %RMSE error.
S9 = 100/(η_{OBS} items number)*SQRT(S8).
S100 = ((10^Q100-10^R100)/(10^Q100))^2.

References

1. Borg, T., Pääkkönen, E. J. (2009). Linear viscoelastic models. Part II. Recovery of the molecular weight distribution using viscosity data. *Journal of Non-Newtonian Fluid Mechanics 156*, 129–138.
2. Bruckner, G., & Cheng, J. (2000). Tikhonov regularization for an integral equation of the first kind with logarithmic kernel. *Journal. Inverse Ill-posed Problems, 8*, 665–675.
3. Borg, T., & Pääkkönen, E. J. (2015). Linear viscoelastic model for different flows based on control theory. *Appl Rheol, 25*, 1–10.
4. Arfken, G., Weber, H. (2001) *Mathematical Methods for Physicists* (5th) (Academic 2001).
5. Borg, T., Pääkkönen, E. J. (2009). Linear viscoelastic models. Part III. Start-up and transient flow effects from the molecular weight distribution. *Journal of Non-Newtonian Fluid Mechanics 159*, 17–25.
6. Meissner, J. (1975). Basic parameters, melt rheology, processing and end-use properties of three similar low-density polyethylene samples. *Pure and Applied Chemistry, 42*, 551–662.
7. Meissner, J. (1972). Modifications of the Weissenberg Rheogoniometer for measurement of transient rheological properties of molten polyethylene under shear. Comparison with tensile data. *Journal of Applied Polymer Science 16*, 2877–2899.
8. Cox, W. P., & Merz, E. H. (1958). Correlation of dynamic and steady state flow viscosities. *Journal of Polymer Science, 28*, 619–622.

Chapter 6
Relaxation Modulus

We have developed a formula for the relaxation modulus $G(t)$ based on control theory principle. This model was originally created to determine the molecular weight distribution (MWD) form viscoelastic measurements. Unlike classical rheological principles, which primarily associate the relaxation modulus or spectrum with specific types of phenomena to address *ill-posed problems*, our approach offers more versatile application. Accurately measuring the relaxation modulus can be challenging. This chapter draws upon articles published [1, 2].

The relaxation modulus is a fundamental concept in classical rheology, as described by Ferry [3] and others in Sect. 1.1. Studies have done further for relaxation modulus. Determining the molecular weight distribution (MWD) or predicting viscoelastic properties requires the relaxation modulus or spectrum to be generated as presented by Tsenoglou [4], Des Cloizeaux [5], Wasserman and Graessley [6], Cassagnau et al. [7], Mavridis and Shroff [8], Mead [9], Wasserman [10], Wasserman and Graessley [11], Nobile et al. [12], Maier et al. [13], Léonardi et al. [14], Thimm et al. [15],Thimm et al. [16], van Ruymbeke et al. [17], Léonardi et al. [18], van Ruymbeke et al. [19], Cocchini and Nobile [20], Vega et al. [21]. To overcome this ill-posed problem, several authors have recently used the double reptation model or the general mixing rule as analysed by Anderssen and Mead [22]. Thimm et al. [15] provided the first analytical relation for deriving the MWD from the modelled relaxation spectrum, even though this is impossible to measure directly. Instead of using the relaxation modulus and spectrum for modelling viscoelastic flows, we use RED function, which is related to all types of flows and polymer structure through MWD.

We introduce two models for the relaxation modulus: The first is an Analytical Model grounded in the impulse function and Control Theory, well-established in mathematics, nuclear physics, and signal control systems. It utilizes only two constants dependent on the polymer's chemical structure, eliminating the need for ad hoc adjustments. The second, the *Unified* (Characteristic) Model, builds on the first model's findings, offering a more practical approach by using the measured

T. Borg, *Control Theory in Rheology*, https://doi.org/10.1007/978-3-031-88248-7_6

time-dependent relaxation modulus related to the MWD, $w(M)$. The computational strategy begins with identifying a form of the MWD and its relations using the rapidly analysed Unified Model, then, if necessary, validating these results with the Analytical Model. This section outlines the new method's main principles, utilizing the familiar relaxation modulus, $G(t)$, and describes the process for simultaneously computing the MWD and the broad-range curve of the linear viscoelastic relaxation modulus from simulated measurements of the relaxation modulus.

6.1 Theory

The formula for the relaxation modulus is developed using the principle of Control Theory. The introduction to shear and complex flows, the relation of the relaxation modulus to molecular structure, is discussed in Chapters 4 and 5. Here, we delve deeper into the molar background.

Modern Control Theory, as applied to dynamic systems, is utilized to model the relaxation modulus. One of the most crucial forms of analysis is input–output modelling, where the output data generated by applying a test input to a system are analysed. This analysis provides valuable insights into cause-effect relationships and helps simplify the model. Given that this perspective is novel in this field, we will elucidate its fundamentals here.

Molecular theories of rheology based on independent chain response (elastic dumb-bell, Rouse, Zimm) or pseudo-independent chain response (unmodified Doi-Edwards) can be used for predicting of the MWD according to Graessley [23] in *Polymeric Liquids and Networks: Dynamics and Rheology* (2008). For such systems, equations for viscosity, recoverable compliance and relaxation time in systems of arbitrary polydispersity can be obtained from the stress relaxation modulus, expressed as the sum of independent contributions. Thus, Graessley wrote Eq. 6.1 as follows:

$$G(t) = \sum w_i G_i(t) \tag{6.1}$$

in which w_i is the weight fraction of component i in the mixture, and $G_i(t)$ is the relaxation modulus for the monodisperse component. We continue to develop the above principle further by converting weight fraction $w_i(M)$ as a function of molecular weight M to the function of time t or effective fraction $w'_i(t) = w_i(M)$ with the found relation between t and M discussed in Chapter 7. A similar conversion, but in the opposite direction, is widely used in wet chemical methods such as gel-permeation chromatography (GPC) tracing with time- or size-exclusion chromatography (SEC) for detecting the MWD.

Knowledge of the factors that influence the effective fraction, $w_i(t)$, is not essential, such as different types of chain dynamics, molecular friction and elasticity, reptation, primitive path fluctuations, constraint release, and other entangled and disentangled chain dynamics.

Chain types in the effective fraction $w_i(t)$ can conform to any linear short- or long-chain branched (SCB and LCB, respectively) structures, combs, networks, H-shapes with multiple arms, or any combinations of complete molecules or their segments. The final structure, molecular weight or construction of the statistical $w_i(t)$ fraction of a single molecule or groups of molecules is not important at this point—only its effects on viscoelasticity. Chapter 7 discusses the real sizes of the fractions. Now we rewrite Eq. 6.1 to get relaxation modulus functional as presented by Anderssen and Loy [24] for $G(t)$ as follows:

$$G(t) = \sum w_i(t)G_i(t) \tag{6.2}$$

We introduce the *rheologically effective distribution* (RED), $w(t)$, and impulse response $h(t)$ after induced stress at time t_0. Relaxation modulus $G_i(t)$ in Eq. 6.2 has relation to a scaled product of $P'G_0h(t_i)$, in which constants G_0 is zero modulus and P' is acting as a scaling factor. Thus we can rewrite Eq. 6.2 for $G(t)$ as

$$G(t) = G_0P' \int_{-\infty}^{t} w(\tau)h(t-\tau)d\tau, \tag{6.3}$$

which can be explained and developed further by the control theory with the explanation starting from the basis. We excite the system with a small stress induced by a small pulsed strain that is applied at time t_0. Pulse response $y(t)$ is obtained from impulse response $h(t)$ and by sampling the active molecules in distribution $w(t)$ between some time interval:

$$y(t) = \int_{-\infty}^{t} w(\tau)h(t-\tau)d\tau \tag{6.4}$$

This equation is the familiar linear formula used in control theory, which is known as a novel principle to adjust and rule one variable or function in a closed system and now in our case for MWD. Since the MWD function is normally a function of logarithmic variables, or here RED $w(\log t)$, we have to rewrite all variables and functions in Eq. 6.4 on a logarithmic scale. The logarithmic convolution (also known as the scale convolution) can be related to ordinary convolution by taking the logarithm of t, and since this is a linear operation it is commutative, associative, and distributive. Thus we can write $h(t) = t$ and $d\tau$ as $d(\log \tau)$. Pulse response $y(\log t) = \log(G(t)/G_0)$ is a normalized relaxation modulus with a maximum value of zero on the logarithmic scale at t_0. The value for zero relaxation modulus $G_0 = G(t_0)$ is obtained by fitting $G(t)$ to experimental measurements. We then obtain the complete relaxation formula in the case where a small and constant strain is induced:

$$\log \frac{G(t)}{G_0} = -P' \int_{-\infty}^{\log t} w(\log \tau)(\log t - \log \tau)d \log \tau \tag{6.5}$$

With true distributions the normalized response on the logarithmic scale decreases, making the overall $G(t)$ a completely monotonically decreasing memory function, as analysed generally by Anderssen and Loy [25], and thus we have to add a negative sign to the right-hand side of Eq. 6.5. Entanglement constant P' is included since not all molecules in the distribution $w(t)$ related to MWD $w(M)$ will be active during the relaxation process due to intermolecule interactions by Matsuoka [26]. Also, the relation $(\Delta G(t)/\Delta w(t)) \propto P'$ at time step Δt depends on the baseline chemical molecular structure, chain types (SCB or LCB) and the average molecular weight. Therefore, it is necessary to adjust the normalized $w(\log t)$ distribution with the value of P' by feedback to obtain the fit for $G(t)$ as illustrated in Fig. 6.1, where the normalized distribution in Eq. 6.5 is

$$\int_{-\infty}^{\infty} w(\log t) d \log t = 1 \tag{6.6}$$

We present normalization procedure in the Sect. 3.8 also for spreadsheet calculations.

RED function $w(t)$ can be regarded as a GPC trace with a time exclusion chromatography or SEC elugram function as a function of time. The largest molecules exit first from the GPC/SEC column, in a similar way to these molecules initially having greater relaxation effects with many entanglements and structural units during relaxation experiments or respectively orientating at the lowest frequency and shear rate. To clarify the differences in the presented principle, we rewrite the simplified

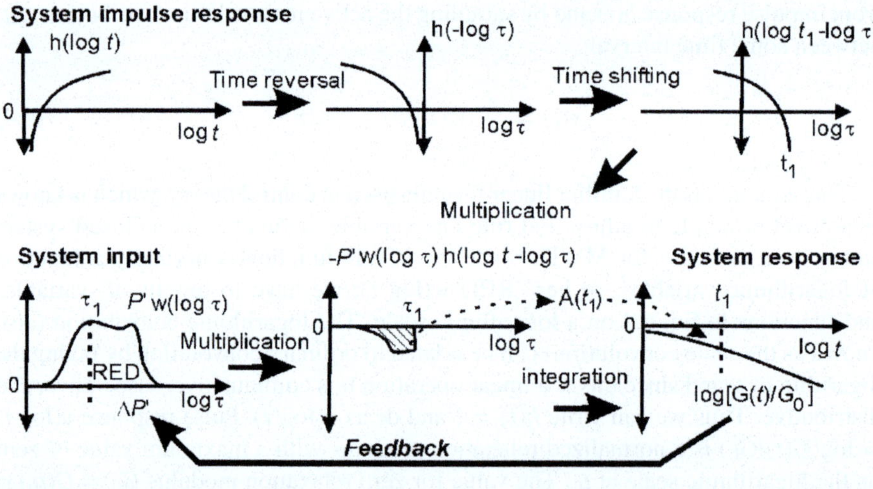

Fig. 6.1 Graphical demonstration of the system and polymer melt. The stress in the molecules decreases during relaxation in distribution $w(\log t)$, and the multiplication product must be negative. The response at time t_1 is defined by area $A(t_1)$, which is scaled by factor P' to obtain the system response as a normalized relaxation function $\log(G(t)/G_0)$. Precise modelling of system response $G(t)$ for the system input based on feedback information P' yields the correct RED function. Borg and Pääkkönen [1]

general form of classical integral equations with kernel $k(t, M)$ used in principle by [4–25], where relaxation time procedures are essential. Thus, for normalized relaxation modulus $g(t) = G(t)/G_0$, we get a Fredholm-type integral equation:

$$g(t) = \int_{-\infty}^{\infty} w(M)k(t, M)dM \tag{6.7}$$

To aid the comparison, we rewrite Eq. 6.3 in a similar simplified style as

$$g(t) = \int_{-\infty}^{t} w(\tau)h(t-\tau)d\tau \tag{6.8}$$

Both Eqs. 6.7 and 6.8 are known as convolution integrals, but the functional Eq. 6.8 has no relaxation time procedures or variables at different scales. Thus there are fundamental differences in relaxation times λ and their discrete spectra are artificial, whereas continuous distribution RED $w(t)$ is a true function in the form of statistical distribution for viscoelastic effects and MWD. The form of the RED $w(\log t)$ function may be roughly similar to the relaxation time spectrum as discussed in Chapter 16. In summary, there are essential and fundamental mathematical differences between the models, and the presented RED function $w(\log t)$ includes all types of chain dynamics.

6.2 Model for Real Polymer Melts by the Analytical Model

Real polymer relaxation occurs in two phases: Rouse [27] relaxation, followed by entangled chain dynamics and reptation, primitive path fluctuations and constraint release. We refer to the second phase here as entanglement relaxation that exhibits mostly elastic effects and responses. Molecular entanglements will relax after deformation roughly at times $t > 0.001$ s, where most observations of $G(t)$ are made. Rouse theory starts from the molecular friction coefficient, where includes rapid elastic dumbbell effects that are mostly relaxed before melts are measured. We employ familiar terminology to introduce the principle of Rouse (viscous) and entanglement (elastic) relaxations, but as discussed in Sect. 6.1, the real chain dynamics are still open and both Rouse and entanglement relaxations have viscous and elastic components. Since we have the same MWD acting for both Rouse relaxation and later for entanglement relaxation, we also use the entanglement distribution in common logarithm scale $w'(\log t)$ for the earlier effective Rouse distribution $w''(\log t_R)$, which corresponds to $w''(\log t - \log R) = w''(\log t/R)$ by standard time shifting. This is achieved simply by copying function $w'(\log t)$ after dividing t by the ratio, R, of the effective or "relaxation time" ranges. The distance for the same point for the effective Rouse and entanglement relaxations and the respective distribution, $w'(\log t)$, is $R = t_e/t_R$. Since Rouse relaxation occurs roughly at $t < 0.001$ s (which is outside the measurement range) and the form of the effective distribution has a minimal effect on the obtained relaxation, the use of $w''(\log t/R)$ with a similar form is the best available choice without more information; nevertheless, the form of the w'' distribution can also differ. We use familiar viscoelastic notation since RED (or RED') $w'(\log t)$ and RED" $w''(\log t/R)$ exhibit mainly elastic and viscous effects, respectively, in the observation range. We convert the relaxation formula of Eq. 6.5 to the entanglement and Rousean relaxation described by Eq. 6.9 and impulse response $\log t - \log \tau = \log(t/\tau)$. In practice, the same distribution, $w'(\log t)$, is copied earlier to the timescale, and we obtain the following complete relaxation formula:

$$\log \frac{G(t)}{G_0} = -\int_{-\infty}^{\log t} \left(P'w'(\log \tau) + P''w''\left(\log \frac{\tau}{R}\right)\right) \log \frac{t}{\tau} d \log \tau \tag{6.9}$$

where the scalar values are $0 < P' < 40$ for entanglement and $0 < P'' < 40$ for the Rouse relaxation ranges. The values of P' and P'' are not ad hoc, but are developed by the software during the procedure of fitting the relaxation modulus data and $w(t)$ distribution. Equation (6.9 models relaxation modulus $G(t)$ without difficulty, but solving distribution $w'(\log t)$ or functions with a logarithmic kernel is a severely ill-posed problem reported by Bruckner and Cheng [28].

Therefore, in Sect. 6.3 we present an alternative solution method.

6.3 Effective Distribution *W'*(*t*) by the Unified Model

We have to extract effective distribution $w'(t)$ from Eq. 6.9, but attempting to do this directly leads to an ill-posed problem. Thus, to make the computations more practical, we use the faster and simpler Unified Model—referred to as the Characteristic Model in a published paper—to obtain the form of the RED curve and MWD. We first approximate the convolution integral in Eq. 6.9 as a summation, from which it is then possible to solve derivative $w'(t)$ directly by computation. Equation 6.4 was previously converted to the logarithmic form for system response $y(\log t)$, which is actually the sum of impulse responses as a series of logarithmic time steps $\Delta\log T$ as a function of i items:

$$y(\log t) = \sum_{i=0}^{\infty} w(i\Delta \log T)\Delta \log T h(\log t - i\Delta \log T) \tag{6.10}$$

If we set $\Delta\log T \to 0$, this equation simplifies to the original convolution form shown in Eqs. 6.4 and 6.5. To obtain the sample or Unified response $y(\log t)$ for Unified (characteristic) relaxation time τ, we set it equivalent to the shifted impulse response, or $i\Delta\log T = \log \tau$. The time-shifted impulse response is now converted to $\log t/\tau$, and the complete formula becomes a standard integral. See illustrations in Fig. 6.1. Thus, in Analytical Eq. 6.9, relaxation modulus $G(t)$ can be approximated as a standard integral of the RED and we obtain Unified relaxation modulus $G(t)$ for constant τ as

$$\log \frac{G(t)}{G_0} = -\log \frac{t}{\tau} \int_{-\infty}^{\log t} \left(P' w'(\log \tau) + P'' w''\left(\log \frac{\tau}{R}\right)\right) d \log \tau \tag{6.11}$$

The logarithmic impulse response has the greatest influence in this special case at time t_0 in Eqs. 6.4 and 6.11, where τ is also rather small, approaching $\Delta \log T$ at small values of index *i*. These points are most representative of the convolution and the relations from *w'*(log *t*) to *G*, which we show in Sect. 6.2 to be within the measurement error. A novel idea is to use originally normalized *G*(*t*) and *w'*(*t*) distribution in both methods, for which the lack of absolute values do not cause difficulties. This procedure removes the absolute value of the pulse response, but this is not a problem since we were originally using only relative values. The relaxation modulus obtained from this simple Unified Model is very close to that obtained from the Analytical Model shown in Fig. 6.3, and within the *G*(*t*) measurement errors. Both the Analytical and Unified Models have the same distribution *w'*(*t*) giving the system response as relaxation modulus data *G*(*t*). Of even more importance is that Eq. 6.12 can be solved for apparent and effective distribution *w'*(*t*) simply by deriving from the measured *G*(*t*) to accurately obtain the shape of the RED curve by Eq. 6.12 and topic discussed more in Sect. 7.4.

$$w'(\log \dot{\gamma}) = -\frac{d}{d \log t} \frac{1}{P'} \left(\frac{\log \frac{G(t)}{G_0}}{\log \frac{t}{\tau_c}} + P'' \int_{-\infty}^{\log t} \left(w''\left(\log \frac{\tau}{R}\right)\right) d \log \tau \right) \tag{6.12}$$

6.4 Converting by Melt Calibration Equations Between Scales

Converting the normalized rheologically effective distribution RED *w*(log *t*) to molecular weight distribution MWD *w* (log *M*) is discussed in the Melt Calibration Chapter 7, but we already present main formulas. For time *t* scale, conversion equations are as follows:

$$w(\log M) = w(\log t)\Big|_{t = t_1 \left(\frac{Mf}{M}\right)^{Hf}} \tag{6.13}$$

$$\tilde{w}(\log t) = w(\log M)\Big|_{M = Mf \left(\frac{t}{t_1}\right)^{-\frac{1}{Hf}}} \tag{6.14}$$

The *Mf* is the polymer structural value that depends on the molecular structure and weight. *Hf* is the conversion factor between scales and is usually 2–4, and *Mf* takes values in the range of 30,000 g/mol for PE.

To maintain consistency among constants and variables, we need to normalize converted RED $\tilde{w}(\log t)$. General normalization of the cumulative distribution for RED $\tilde{w}(\log t)$ can be done by Eq. 6.6 and procedure is discussed in the Sect. 3.8 also for spreadsheet calculations.

To get differential Molecular Weight Distribution from MWD $\tilde{w}(\log M)$ in the form normally presented $dW/d(\log M)$ is explained in Sect. 3.4.

6.5 Constants for Modelling

For the Rouse shear range distribution, the default value of R was set to 10^6, except for Fig. 6.2, where $R = 10^4$ was used for descriptive reasons. For the LDPE sample, $Mf = 37{,}800$ g/mol at 150 °C was used with the year 2005 measured MWD data. The scale exponent for PE, $Hf = 2.05$, was kept constant for polyolefins tested with a plate–plate oscillating rheometer. It's important to note that using a cone-plate head ($Hf = 2.5$) would result in a gap between plates that is too narrow for very long PE molecules.

For comparison, constants for PS were $Mf = 136{,}000$ g/mol and $Hf = 4$, as commonly used for all other polymers except for polyolefins mentioned earlier. P' and P'' values were developed through the $G(t)$ fitting procedure, and no ad hoc constants or values were used. A constant characteristic time of $\tau_c = 10^{-6}$ s was employed in the relaxation simulations.

a. The normal or Gaussian distribution as rheologically effective distribution (RED), $w(t)$, plotted on a logarithmic scale with $\tau_c = 1$ s. Its Unified relaxation is very close to the classical Maxwellian single-element relaxation, with $\lambda = 100$ s (dashed line).

b. Relaxation by Eq. (6.11) using the Weisslau distributions $w(\log t)$ and $w(\log t/R)$, with constants $\tau_c = 10^{-6}$ s, P' and P'' set to 0.23, and $R = 10^4$. Both normalized relaxation moduli $G(t)/G_0$ and logarithmic $\log G(t)$ with illustrated $G_0 = 10$ Pa are shown. A typical similar real polymer has the average molecular weight Mw in the

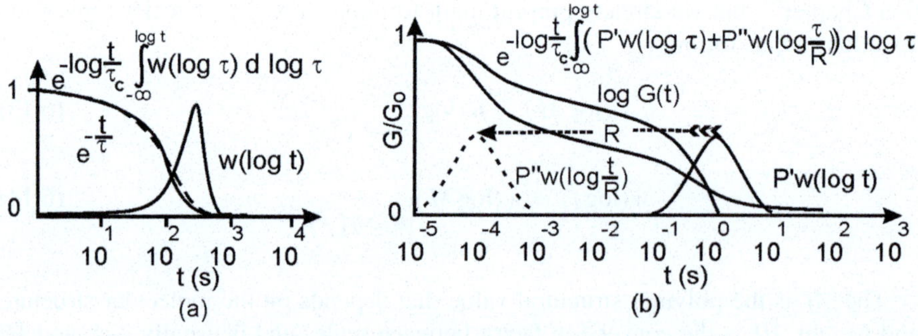

Fig. 6.2 Schematic models of the relaxation modulus

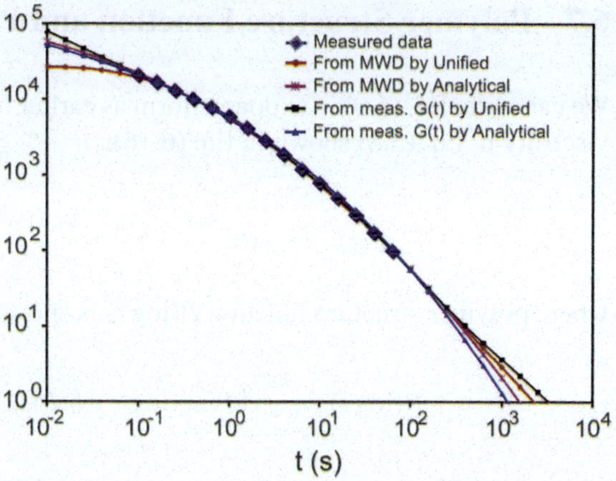

Fig. 6.3 Relaxation modulus for IUPAC A LDPE in the range of measurements performed by Meissner, from 0.1 s only up to 80 s at 150^{0}C. Both Analytical Eq. (6.9) and the Unified (Characteristic) Eq. (6.11) formulas are used to show minimal differences between models in the measurable range. Borg and Pääkkönen [1]

range of 100,000g/mol, and the polydispersity index, *MwR*, is below 1.2 in relaxation processes. Borg and Pääkkönen [1].

Figure 6.2a shows that Unified Model with normal distribution according to Eq. (3.1) with mean μ = -100, variance $\sigma^2 = 225$ and multiplied value 229 models close classical Maxwellian single-element relaxation, with $\lambda = 100$ s.

Figure 6.2b uses respectively standard Wesslau distributions.

6.6 Initial Relaxation Modulus Value from Viscosity Data

In cases where we are using only viscosity data, we do not obtain the absolute level of the relaxation modulus directly. Since we simultaneously obtain the zero viscosity η_0 using a wide-range viscosity fit and the normalized relaxation modulus, $g(t) = G(t)/G_0$, we need tools to find an absolute value for G_0. Here we can use standard simple numerical fitting procedures such as bracketing, while solving Eq. (6.15) by testing with new $\hat{G}_0$ values to obtain the new $G_0 = \hat{G}_0$.

$$\eta_0 = \int_{-\infty}^{\infty} \hat{G} g(t) d \log t \tag{6.15}$$

An alternative, simpler procedure can be used if we know even a single data point in the measured relaxation modulus curve $G(t)$ and MWD with known constants.

6.7 Polymer Structure Function and Power Forms

We can write Eq. (6.9) in exponent form as earlier for shear in Eq. (4.7) and complex viscosity in Eq. (5.8) shown in Eq. (6.16):

$$G(t) = G_0 10^{-\log\frac{t}{\tau}\int_{-\infty}^{\log t}\left(P'w'(\log\tau)+P''w''\left(\log\frac{\tau}{R}\right)\right)d\log\tau} \tag{6.16}$$

where polymer structure function $P(\log t)$ is shown in Eq. (6.17):

$$P(\log t) = \int_{-\infty}^{\log t}\left(P'w'(\log\tau) + P''w''\left(\log\frac{\tau}{R}\right)\right)d\log\tau \tag{6.17}$$

where we can get elastic $P'(\log t)$ and viscous $P''(\log t)$ components similar way as in Sect. 4.9 and 5.3.

This gives us a simple power-form formula to Eq. (6.16) included relation $-\log\frac{t}{\tau_c}$ and approximates the true molecular orientation level and state in Eq. (6.18).

$$G(t) = G_0 10^{-\log\frac{t}{\tau_c}P(t)} = G_0\left(10^{\log\frac{t}{\tau_c}}\right)^{-P(t)} \tag{6.18}$$

As we used common logarithm, we can write directly Eq. (6.19).

$$G(t) = G_0\left(\frac{t}{\tau_c}\right)^{-P(t)} \tag{6.19}$$

6.8 Modelling the Relaxation Modulus

We aimed to utilize relaxation data measured at 150 °C for IUPAC A, as manually obtained from the sources cited by Meissner [28, 29]. In RheoDeveloper (RheoPower), we sketched an MWD curve by eye to closely match the original relaxation modulus curve. The originally measured average molecular weight *Mw* for this LDPE was 472,000, with a polydispersity index, *MwR*, of 24.9 year 1975. Therefore, we employed *Mw* = 480,000 and *MwR* = 25 in our first modelling attempts. However, as we have shown, more important is the used RED converted from MWD, as RED remains largely consistent over the years and independent of the measured MWD by GPC/SEC over the decades, as discussed in Sect. 3.8.

A more detailed discussion on the modelling of the relaxation modulus is provided in Sect. 7.4. Simulations for the relaxation modulus are performed, and some results are presented.

In principle, relaxation modulus *G*(*t*) data are suitable as a data source, but have been found to be of limited use in computations due to distortion resulting from procedural and instrumental sources. During experiments, some deformation always occurs before relaxation. The simulation is at best a two-step process, because earlier states and deformation histories considerably influence the measured properties. Inertial forces of the oscillating head of the device also cause visible errors in the $G(t)$ data. However, very recently there have been promising developments in rotational rheometers to obtain more accurate $G(t)$ data.

6.9 Relaxation Modulus and Spectrum at Different Temperatures

The computed relaxation modulus, $G(t)$, is not affected by terminal or instrumentation limitations. We set the maximal relaxation time for the spectrum, λ max, to 400 s.

The relaxation spectrum, $h(\lambda)$, can be computed by deriving the relaxation modulus $G(t)$ following the procedures outlined by Tschoegl

[31]. He utilized the natural logarithmic form. The following equation demonstrates how to derive $h(\lambda)$ using up to the third derivative of $G(t)$:

$$h(\lambda) = -\frac{dG(t)}{d\ln t} + \frac{3d^2G(t)}{2(d\ln t)^2} - \frac{d^3G(t)}{2(d\ln t)^3}\bigg|_{t=3\lambda} \tag{6.20}$$

The wide-range relaxation moduli and relaxation spectra curves for Lupolen 1840H LDPE computed at 13 temperatures are presented in Figs. 6.3 and 6.4, respectively. These curves exhibit smooth behaviour, with only slight variations due to the sensitivity of the derivation.

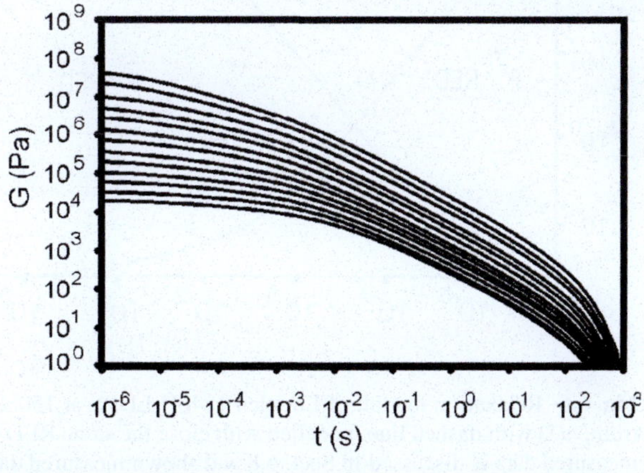

Fig. 6.4 Relaxation moduli of Lupolen 1840H LDPE at temperatures ranging from 130°C to 250°C (in 10°C increments). Borg and Pääkkönen [30]

6.9.1 Practical Examples

To generate an approximation for relaxation modulus $G(t)$. We use a normalized Wesslau distribution for $w(\log t)$ RED simulating Lupolen 1840H LDPE polymer and RED $w(\log t)$ form shown in the Fig. 6.5 and 6.6.

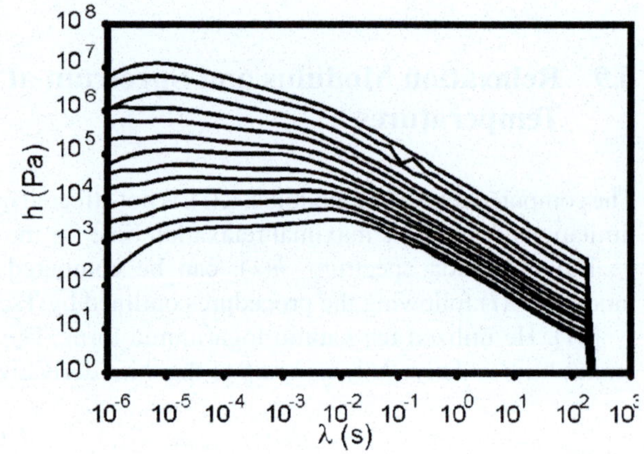

Fig. 6.5 Relaxation spectra of Lupolen 1840H LDPE at temperatures ranging from 130°C (top curve) to 250°C (bottom curve) (in 10°C increments). Borg and Pääkkönen [30]

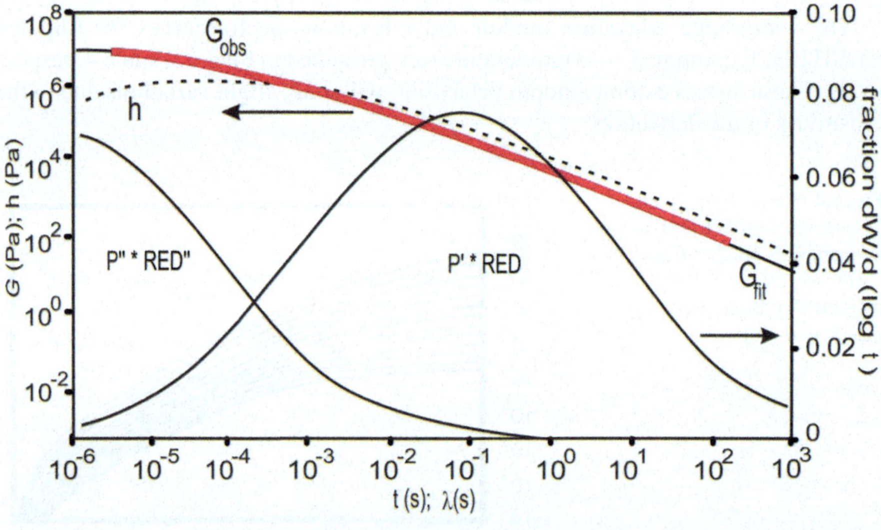

Fig. 6.6 Relaxation moduli of Lupolen 1840H LDPE at 150°C temperature and relaxation spectrum, $h(\lambda)$ with dashed line modelled with close the same RED by Wesslau distribution. Source of measured data is discussed in Sect. 6.8 and shown measured data is a composition by RheoPower software, as shown in the thick segment of the curve. As constant characteristic time of $\tau_c = 10^{-6}$ s was employed in the relaxation simulations, to model smaller time ranges is not possible due to completely monotone fading memory property for relaxation modulus by Anderssen and Loy [25]

6.9.2 Example on a Spreadsheet

We model the relaxation modulus measurements of 1840H LDPE using artificial Wesslau distributions for RED, as shown in Fig. 6.6, with closely matching average molecular weight and polydispersity values to those of 1840H LDPE. Undoubtedly, measured MWD for 1840H LDPE can be utilized, but this data sample would require 200 rows for display in subsequent spreadsheets.

For Wesslau distribution is found the best fit viscosity by values $\mu =$ -0.9 and $\sigma =$ 1.92 to those modelled from Meissner and Laun's measurements. The elastic value $P' = 0.36$ and viscous $P'' = 0.30$. By fitting with measurements is found relaxation modulus constant $G_0 = 10^{6.84}$ Fig. 6.11.

6.9.3 Formulas on the Spreadsheet for Relaxation Modulus

Attached in Fig. 6.7 is cut of pre-prepared calculation spreadsheet with formulas of the same computation as get Fig. 6.6. In the Sect. 3.11 is explained some general principles for presented spreadsheets. Figure 6.8 shows same cells in the Formula Auditing Mode Figs. 6.9 and 6.10.

Input values:

C5 −0.9 (μ is the mean).

	B	C	D	E	F	G	H	I	J	K	L	M	N	O	P	Q	R	S
3	RELAXATION MODULUS AND SPECTRUM																	
4																		
5	μ	-0.9		R'		0	P'	0.36		%RMSE		0.11	%					
6	σ	1.92		R''		5	P''	0.3		log		τ	-6	0.000001				
7							log G/o	6.84										
8		x=time (s)							1.00					0.99				0.01
9	row	x	log x	w	log x'	tilde w(log x)	RED' w(log x)	int (log x')	P' (log x')	log x''	tilde w''(	RED''	int (log x''	P'' (log x'')	P(x)	log G(t)	Measured	0.11
10	0	0.00	-10	2.75E-06	-10	2.75226E-06	9.90813E-07	0.00	0.00	-5.00	0.02	0.0071	0	0.00	8.05E-04	6.84		
11	1	0.00	-9.9	3.52E-06	-9.9	3.51809E-06	1.29062E-06	0.00	0.00	-4.90	0.03	0.0079	0.00	0.00	8.05E-04	6.84		

Fig. 6.7 Presents a top section of an MS Excel 365 spreadsheet used for modelling relaxation modulus. Cells outlined with thick borders and containing bold numbers are designated for input data. The RheoPower software internally determines these values when computing viscosity measurements data. Cells framed by thinner borders represent output data or are intended solely for verification purposes

	B	C	D	E	F	G	H	I	J
3	RELA								
4									
5	μ	-0.9		R'		0	P'	0.36	
6	σ	1.92		R''		5	P''	0.3	
7							log G/o	6.84	
8		x=time (s)							=+I210
9	row	x	log x	w	log x'	tilde w(log x)	RED' w(log x)	int (log x')	P' (log x')
10	0	=10^D10	-10	=1/($C$6*SQRT(2*PI()))*EXP(-1/2*((LOG10(C10)-$C$5)/$C$6)^2)	=+D10+G$5	=XLOOKUP(F10;D$10:D$210;E$10:E$210;0;1;1	=+G10*I$5	=+G10	=+I10/J$8*I$5
11	=+B10+1	=10^D11	=+D10+0.1	=1/($C$6*SQRT(2*PI()))*EXP(-1/2*((LOG10(C11)-$C$5)/$C$6)^2)	=+D11+G$5	=XLOOKUP(F11;D$10:D$210;E$10:E$210;0;1;1	=+AVERAGE(G10:G12)*I$5	=G11*(D11-D10)+I10	=+I11/J$8*I$5

Fig. 6.8 Left part of the formulas for modelling relaxation modulus and component P'. Cells outlined with thick borders and containing bold numbers are designated for input data. Cells framed by thinner borders represent output data or are intended solely for verification purposes

	K	L	M	N	O	P	Q	R	S
3									
4									
5	%RMSE		=+S9	%					
6	log		τ	-6	=10^N6				
7									
8					=+N210				=+SUM(T120:T127)
9	log x"	tilde w"(log x)	RED"	int (log x")	P" (log x")	P(x)	log G(t)	Measured	=100/107*SQRT(S8)
10	=+D10+G$6	=XLOOKUP(K10;D$10:D$210;E$10:E$210;0;1;1	=+L10*I$6	0	=+O11	=+O10+J10	=+Q11		
11	=+D11+G$6	=XLOOKUP(K11;D$10:D$210;E$10:E$210;0;1;1	=+L11*I$6	=L11*(D11-D10)+N10	=+N11/O$8*I$6	=+O11+J11	=+LOG(POWER(10;I$7)*POWER(C11/O$6;-P11))		

Fig. 6.9 Right side part of the formulas for modelling and component P'' and complete relaxation modulus. Columns R and S with row 11 is demonstrating data and formulas for comparing measured and observed with modelled viscosity

	U	V	W	X	Y
3	RELAXATION SPECTRUM				
4					
5					
6					
7					
8			Derivations		
9	G(t)	h(lambda)	First	Second	Third
10	6968494				
11	6968494				

Fig. 6.10 Presents a top section of an MS Excel 365 spreadsheet for modelling relaxation spectrum $h(\lambda)$ according to Eq. 6.15 from $G(t)$. Since several derivations are required, it is easiest to perform them in different columns

	U	V	W	X	Y
60	=10^Q60	=LOG10(-W60+X60-Y60)	=(U61-U60)/(LN(C60)-LN(C59))	=3*(W61-W60)/(2*(LN(C60)-LN(C59)))	=(X61-X60)/(2*(LN(C60)-LN(C59)))
61	=10^Q61	=LOG10(-W61+X61-Y61)	=(U62-U61)/(LN(C61)-LN(C60))	=3*(W62-W61)/(2*(LN(C61)-LN(C60)))	=(X62-X61)/(2*(LN(C61)-LN(C60)))
62	=10^Q62	=LOG10(-W62+X62-Y62)	=(U63-U62)/(LN(C62)-LN(C61))	=3*(W63-W62)/(2*(LN(C62)-LN(C61)))	=(X63-X62)/(2*(LN(C62)-LN(C61)))
63	=10^Q63	=LOG10(-W63+X63-Y63)	=(U64-U63)/(LN(C63)-LN(C62))	=3*(W64-W63)/(2*(LN(C63)-LN(C62)))	=(X64-X63)/(2*(LN(C63)-LN(C62)))

Fig. 6.11 Presents lines from 60 to 63 to calculate relaxation spectrum $h(\lambda)$ according to Eq. 6.20 from $G(t)$. Please use the natural logarithmic mode for derivations. Since calculations cannot be made at very small relaxation times, lines 60 to 63 were selected to represent time in the range of log − 4

G5 0 (*R'* the place of RED' at a rate scale for copy of RED).
I5 0.36 (*P'* elastic value).
C6 1.92 (σ is standard deviation).
G6 5 (log *R*, *R*" the place of RED" at a rate scale for copy of RED).
I6 0.30 (*P*" viscosity value).
O6 −6 (log τ_0 constant characteristic time).
I7 6.84 (log zero modulus G_0 at τ_0).
D10 −10 (initial log rate for rows and arrays).

Formulas:

O6 = 10^N6.
G8 = + SUM(G10:G210).
I8 = + SUM(I10:I210).

J8 = + SUM(J10:J210).
J8 = + I210.
B11 = + B10 + 1.
C11 = 10^D11.
D11 = + D10 + 0.1
E11 = 1/($C$6*SQRT(2*PI()))*EXP(-1/2*((LOG10(C11)-$C$5)/$C$6)^2).
F11 = + D11 + G$5.
G11 = XLOOKUP(F11;D$10:D$210;E$10:E$210;0;1;1).
H11 = + AVERAGE(G10:G12)*I$5.
I11 = G11*(D11-D10) + I10.
J11 = + I11/J$8*I$5.

Formulas:

O8 = + N210.
K11 = + D11 + G$6.
L11 = XLOOKUP(K11;D$10:D$210;E$10:E$210;0;1;1).
M11 = + L11*I$6.
N11 = L11*(D11-D10) + N10.
O11 = + N11/O$8*I$6.
P11 = + O11 + J11.
Q11 = + LOG(POWER(10;I$7)*POWER(C11/O$6;-P11)).

Option input and formulas:

M5 = + S9.
S8 = + SUM(Sa:Sb) first row Sa and Sb of η_{OBS} data for %RMSE error.
S9 = 100/(η_{OBS} items number)*SQRT(U8).
S100 = ((10^Q100-10^R100)/(10^Q100))^2.

6.9.4 Formulas on the Spreadsheet for Relaxation Spectrum

Formulas:

U60 = 10^Q60.
V60 = LOG10(-W60 + X60-Y60).
W60 = (U61-U60)/(LN(C60)-LN(C59)).
X60 = 3*(W61-W60)/(2*(LN(C60)-LN(C59))).
Y60 = (X61-X60)/(2*(LN(C60)-LN(C59))).

References

1. Borg, T., Pääkkönen, E. J. (2009). Linear viscoelastic models. Part I. Relaxation modulus and melt calibration. *Journal of Non-Newtonian Fluid Mechanics 156*, 121–128.

2. Borg, T., Pääkkönen, E. J. (2010) Linear viscoelastic models. Part IV. From molecular dynamics to temperature and viscoelastic relations using control theory. *Journal of Non-Newtonian Fluid Mechanics 165,* 24–31.
3. Ferry, J. D. (1980). *Viscoelastic Properties of Polymers* (3rd ed.). Wiley-Interscience.
4. Tsenoglou, C. (1987). Viscoelasticity of binary homopolymer blends. *ACS Polymer Preprint, 28*, 185–186.
5. Des Cloizeaux, J. (1988). Double reptation vs simple reptation in polymer melts. *Europhysics Letters 5*, 437–442.
6. Wasserman, S. H., & Graessley, W. W. (1992). Effects of polydispersity on linear viscoelasticity in entangled polymer melts. *Journal of Rheology, 36*, 543–572.
7. Cassagnau, P., Montfort, J. P., Marin, G., & Monge, P. (1993). Rheology of polydisperse polymers; relationship between intermolecular interactions and molecular weight distribution. *Rheologica Acta, 32*, 156–167.
8. Mavridis, H., & Shroff, R. J. (1993). Appraisal of a molecular weight distribution-to-rheology conversion scheme for linear polyethylenes. *Journal of Applied Polymer Science, 49*, 299–318.
9. Mead, D. W. (1994). Determination of molecular weight distributions of linear flexible polymers from linear viscoelastic material functions. *Journal of Rheology, 38*, 1797–1827.
10. Wasserman, S. H. (1995). Calculating the molecular weight distribution from linear viscoelastic response of polymer melts. *Journal of Rheology, 39*, 601–625.
11. Wasserman, S. H., & Graessley, W. W. (1996). Prediction of linear viscoelastic response for entangled polyolefin melts from molecular weight distribution. *Polymer Engineering Science, 36*, 852–861.
12. Nobile, M. R., Cocchini, F., & Lawler, J. V. (1996). On the stability of molecular weight distributions as computed from the flow curves of polymer melts. *Journal of Rheology, 40*, 363–382.
13. Maier, D., Eckstein, A., Friedrich, C., & Honerkamp, J. (1998). Evaluation of models combining rheological data with the molecular weight distribution. *Journal of Rheology, 42*, 1153–1173.
14. Léonardi, F., Allal, A., & Marin, G. (1998). Determination of the molecular weight distribution of linear polymers by inversion of a blending law on complex viscosities. *Rheologica Acta, 37*, 199–213.
15. Thimm, W., Friedrich, C., Marth, M., & Honerkamp, J. (1999). An analytical relation between relaxation time spectrum and molecular weight distribution. *Journal of Rheology, 43*, 1663–1672.
16. Thimm, W., Friedrich, C., Marth, M., & Honerkamp, J. (2000). On the Rouse spectrum and the determination of the molecular weight distribution from rheological data. *Journal of Rheology, 44*, 429–438.
17. van Ruymbeke, E., Keunings, R., & Bailly, C. (2002). Determination of the molecular weight distribution of entangled linear polymers from linear viscoelasticity data. *Journal of Non-Newtonian Fluid Mechanics, 105*, 153–175.
18. Léonardi, F., Allal, A., & Marin, G. (2002). Molecular weight distribution from viscoelastic data: The importance of tube renewal and rouse modes. *Journal of Rheology, 46*, 209–224.
19. van Ruymbeke, E., Keunings, R., Stéphenne, V., Hagenaars, A., & Bailly, C. (2002). Evaluation of reptation models for predicting the linear viscoelastic properties of entangled linear polymers. *Macromolecules, 35*, 2689–2699.
20. Cocchini, F., & Nobile, M. R. (2003). Constrained inversion of rheological data to molecular weight distribution for polymer melts. *Rheologica Acta, 42*, 232–242.
21. Vega, J.F., Rastogi, S., Peters, G. W. M., Meijer, H. E. H. (2004). Rheology and reptation of linear polymers. Ultrahigh molecular weight chain dynamics in the melt. *Journal of Rheologica Acta 48*, 663–678.
22. Anderssen, R. S., & Mead, D. W. (1998). Theoretical derivation of molecular weight scaling for rheological parameters. *Journal of Non-Newtonian Fluid Mechanics, 76*, 299–306.
23. Graessley, W. W. (2008). *Polymeric Liquids and Networks: Dynamics and Rheology*. Garland Science.

24. Anderssen, R. S., & Loy, R. J. (2001). On the scaling of molecular weight distribution functionals. *Journal of Rheology, 45*, 891–901.
25. Anderssen, R. S., & Loy, R. J. (2002). Rheological implications of completely monotone fading memory. *Journal of Rheology, 46*, 1459–1472.
26. Matsuoka, S. (1992). *Relaxation Phenomena in Polymers*. Hanser.
27. Rouse, P. R., Jr. (1953). A theory of the linear viscoelastic properties of dilute solutions of coiling polymers. *J. of Chem. Phys., 21*, 1272–1280.
28. Meissner, J. (1975). Basic parameters, melt rheology, processing and end-use properties of three similar low-density polyethylene samples. *Pure and Applied Chemistry, 42*, 551–662.
29. Meissner, J. (1972). Modifications of the Weissenberg Rheogoniometer for measurement of transient rheological properties of molten polyethylene under shear. Comparison with tensile data. *Journal of Applied Polymer Science 16*, 2877–2899
30. Borg, T., Pääkkönen, E. J. (2009). Linear viscoelastic models. Part II. Recovery of the molecular weight distribution using viscosity data. *Journal of Non-Newtonian Fluid Mechanics 156*, 129–138.
31. Tschoegl, N. W. (1989). *The phenomenological theory of linear viscoelastic behavior: An introduction* (p. 171). Springler-Verlag.

Chapter 7
Converting Red Function to MWD

The formulas presented here enable the application of the Control Theory (CT) principle. Melt Calibration is in principle Rheological-Molecular Weight Distribution Transformation or Rheological-Molecular Interconversion, but the practical Unified equations and applications start from Sect. 7.3 onward [1].

7.1 Melt Calibration

We need to convert distribution scales between molecular weight scale M and rheologically effective scale as a shear or frequency rate or time. As relaxation procedures are better known in rheology, we use a function of time t by introducing the melt calibration, which is the relation between rheological properties and the molecular weight.

A test point in a relaxed polymer melt after small deformation at t_0 can be approximated statistically as a nonrelaxed sphere of volume V_0 with an average molecular size up to $M(t_0)$. The apparent sphere of effective molecules starts to shrink as a function of radius $r(t)$, with the molecules located farthest from the test point relaxing first. Molecules that are farther from the test point deform less, as shown in the Fig. 7.2a. The average molecular size inside the sphere is $M = Mf\ \mathrm{d}V/V_0$, where Mf is the polymer structural value that depends on the molecular structure and weight. The effective volume for the nonrelaxed sphere and thus the effective molecular size change dM, as a function of $r(t)$, is given by $\mathrm{d}M = Mf\frac{4}{3}\pi\frac{\mathrm{d}r^3}{V_0}$, where $Hf = \frac{4}{3}\pi \approx 4.19$ is the conversion factor between scales M and t. By induction, we can similarly model all the test points of the melt, where the distance to walls is greater as $r(t_0)$.

The relation between molecular weight scale M and rigid time scale t is obtained using a homogeneous linear differential formula. The additional decrease in dM converted by Hf on the M scale according to $-\mathrm{d}M/\mathrm{d}t$ must equal the molecular

T. Borg, *Control Theory in Rheology*, https://doi.org/10.1007/978-3-031-88248-7_7

weight scale divided by time, M/t, or

$$Hf\frac{\mathrm{d}M}{\mathrm{d}t} + \frac{M}{t} = 0. \tag{7.1}$$

7.2 RED and MWD are Mirrored Functions

Solving Eq. (7.1) yields a simple relation for $t > 0$, which is the melt calibration, $M(t)$, as a function of time, where for the factor Mf value is M at $t_1 = 1$ s:

$$t = t_1\left(\frac{Mf}{M}\right)^{Hf} \tag{7.2}$$

Setting Eq. (7.2) to RED function $w(\log t)$ we get function composition function as follows for MWD $w(\log M)$:

$$w(\log M) = w\left(t_1\left(\frac{Mf}{M}\right)^{Hf}\right) \tag{7.3}$$

Before going further few words about functions composites, which is an operation $\circ$ that takes two functions f and g, and produces a function $h = g \circ f$ such that $h(x) = g(f(x))$. A used example is if an airplane's altitude at time t is $a(t)$, and the air pressure at altitude x is $p(x)$, then $(p \circ a)(t)$ is the pressure around the plane at time t.

Non-normalized $\tilde{w}(\log M)$ is now mirrored distribution for vertical w-axes and RED function $w(\log t)$. Well known statistical Gumbel distribution [2] has inverted (mirrored) distribution, but in our case the abscissas are also different. Due to logarithmic scales normalised MWD $w(\log M)$ and RED $w(\log t)$ have in principle similar distribution forms, only mirrored to each other. Depending on the used scales the width of distribution functions can be different (see Fig. 7.1).

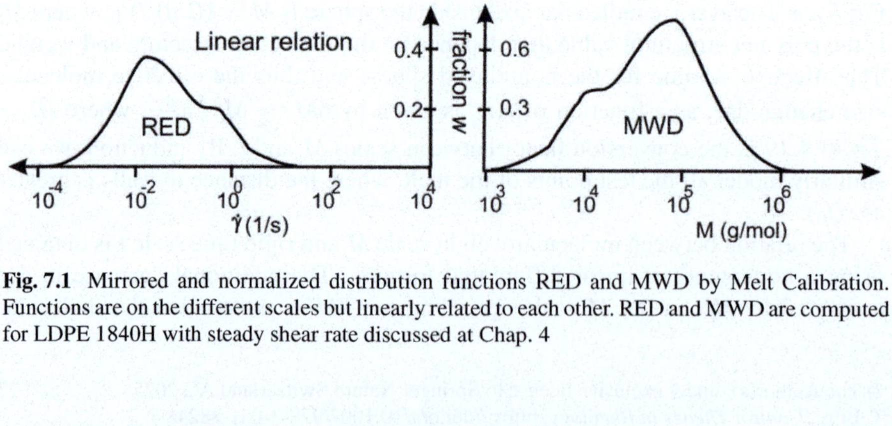

Fig. 7.1 Mirrored and normalized distribution functions RED and MWD by Melt Calibration. Functions are on the different scales but linearly related to each other. RED and MWD are computed for LDPE 1840H with steady shear rate discussed at Chap. 4

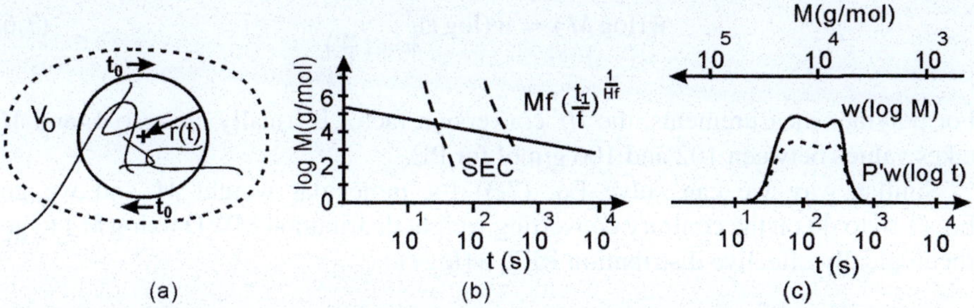

Fig. 7.2 Relation model for melts, and conversions from the RED to the MWD. **a** In a relaxed polymer, a test point in the melt after a small shear deformation at t_0 is equivalent to a statistical nonrelaxed sphere of volume V_0, which shrinks as a function of radius $r(t)$ as the farthest molecules or the ends of individual molecules relax first, resulting in less deformation relative to the test point. **b** Melt calibration curve $M(t)$ for PS, and the respective typical universal calibration curves for different SEC columns marked by dashed lines. **c** RED $w(\log t)$ from the time scale converted to inverse molecular weight scale w (top) and MWD $w(\log M)$ or $w'(\log M) = w(\log t)$ using Eq. (7.3) or Eq. (7.5). Borg and Pääkkönen [1]

Solving Eq. (7.1) yields a simple relation for t > 0, which is the melt calibration, $M(t)$, as a function of time, where for the factor Mf value is M at $t_1 = 1$ s:

$$M = Mf\left(\frac{t}{t_1}\right)^{-\frac{1}{Hf}} \tag{7.4}$$

We write respective composite function for to RED function $w(\log t)$ from MWD $w(\log M)$ we get as follows:

$$\tilde{w}(\log t) = w\left(Mf\left(\frac{t}{t_1}\right)^{-\frac{1}{Hf}}\right) \tag{7.5}$$

RED function $\tilde{w}(\log t)$ is good to normalize before using as presented in the Sect. 3.8.

7.3 Other Forms for RED and MWD Interconversion Formulas

We present same formulas presented Sect. 7.2 by equations of conditional substitutions, which may be more familiar.

Converting rheologically effective distribution RED $w(\log t)$ to molecular non normalized weight distribution MWD $\tilde{w}(\log M)$ we get formula or

$$\tilde{w}(\log M) = w(\log t)\Big|_{t=t_1\left(\frac{Mf}{M}\right)^{Hf}} \tag{7.6}$$

For polymer measurements, the *Hf* conversion factor is usually close to 4, and *Mf* takes values between 102 and 106 g/mol for PE.

Similar way we can solve Eq. (7.3) for molecular weight M and we get Eq. (7.6) to do on the contrary converting weight distribution MWD $w(\log M)$ to get rheologically effective distribution RED $w(\log t)$.

$$\tilde{w}(\log t) = w(\log M)\Big|_{M=Mf\left(\frac{t}{t_1}\right)^{-\frac{1}{Hf}}} \tag{7.7}$$

Since a common logarithm is generally used for molecular weight M in the $w(\log M)$ MWD, there are practical benefits in using a for the $w(\log t)$ normalized distribution, which we call the *rheologically effective distribution* RED. The form of distribution $w(\log t)$ is in many cases close Wesslau (on non-logarithmic scale as Gaussian) form and can be used of modelling rough viscosity values for practical purposes.

It is good practise to normalize MWD $\tilde{w}(\log M)$ distribution for presentations, but RED $\tilde{w}(\log t)$ after conversion must always normalized. This keep all constants and variables comparative and relative to each other.

Normalization of RED distributions is explained in the Sect. 3.8 also for spreadsheet calculations. To get differential Molecular Weight Distribution from MWD $\tilde{w}(\log M)$ in the form normally presented $dW/d(\log M)$ is explained in Sect. 3.4.

7.4 Relation to GPC/SEC

Benoit et al. [3] introduced the universal calibration concept for GPC (that was subsequently also applied to SEC), which used the hydrodynamic volume, and here the nonrelaxed melt volume is used in the Figs. 7.2a and 7.2b shows the melt calibration curve for polystyrene (PS) and compares it to the SEC calibration.

We need to convert RED $w(t)$ to molecular weight distribution $w(M)$ or in logarithmic scales $w\prime(\log M) = w(\log t)$. When an effective distribution function $w(\log t)$ is found from the best fit to the relaxation modulus, we can convert it by variable transformation to $w(\log M)$ using the standard multivariate change-of-variables formula, Eq. (7.3), and the exponent (i.e., conversion factor between scales *Hf* and the polymer structural value *Mf* results shown in Fig. 7.2c).

The correct value of the *Mf* factor for each polymer type is found from fitting GPC, SEC, absolute multiangle light-scattering measurements, dynamic measurements, and models.

The wide scales of the relaxation modulus curve at 150 °C computed using the integration formula and the MWD are shown in Fig. 7.3a.

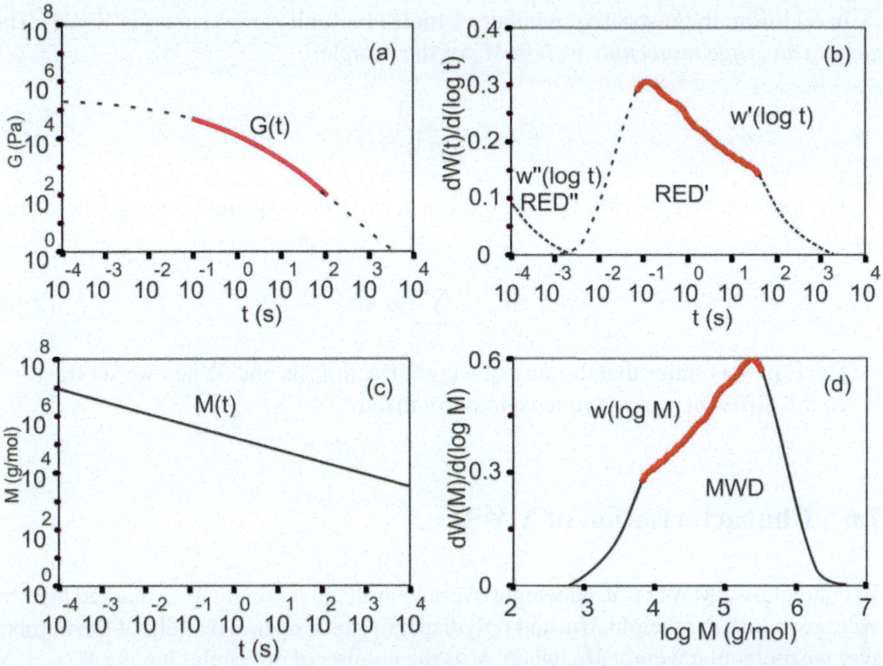

Fig. 7.3 From relaxation modulus to MWD of IUPAC A LDPE at 150 °C. **a** Thicker segment of measured relaxation modulus $G(t)$ and wider thin dashed curve generated by modelling from complete $w(t)$ of Fig. 7.2b using Eq. (6.11). Thicker segment of the curve of the rheologically effective distribution (RED), $w(t)$, was computed directly from measured relaxation modulus $G(t)$ using Eq. (9.5), and thinner dashed segment was obtained by applying a standard best-fit procedure between measured and modelled $G(t)$. Melt calibration curve $M(t)$ according to Eq. (6.12) transforms $w(t)$ to MWD $w(\log M)$. The MWD is converted from RED function in a manner similar to GPC/SEC techniques. It is now possible to analytical compute $G(t)$ backwards from the obtained MWD so as to check the accuracy of the analysis results. Borg and Pääkkönen [1]

The universal calibration concept for GPC/SEC works as a function of time but Melt Calibration principle is suitable for different type flows and we have shown respective Melt Calibration formulas with the corresponding topics of flows.

7.5 Molecular Weight Averages

Let M_i (i = 1, 2, …) be the set of molecular weights represented in a sample, and W_i their respective weight fractions. The sum of weight fractions must be unity:

$$\sum W_i = 1 \tag{7.8}$$

In addition, the respective number of moles per unit sample mass is W_i/M_i. The *number-average molecular weight* M_n of the sample:

$$M_n = \frac{1}{\sum (W_i/M_i)} \tag{7.9}$$

The larger molecules are weighted more heavily in the weight-average molecular weight M_w:

$$M_w = \sum W_i M_i \tag{7.10}$$

The Eq. (3.1) states that the sum of weight fractions is one. When we set fractions W_i to the different scales, we have to normalized.

7.6 Characterisation of MWD

To characterise MWD is used weight average molecular weight $\overline{M}_W$ marked here by average molecular weight *Mw* and polydispersity index *MwR* by help of the number average molecular weight M_n, where N_i is the number of molecules having M_i weight in the polymer sample, and M_i is the weight of a particular molecule of the sample.

$$M_n = \frac{\sum_{i=1}^{n} w_i}{\sum_{i=1}^{n} w_i/M_i} \tag{7.11}$$

$$Mw = \frac{\sum_{i=1}^{n} w_i M_i}{\sum_{i=1}^{n} w_i} \tag{7.12}$$

$$MwR = \frac{Mw}{M_n} \tag{7.13}$$

On the semi-logarithmic charts for molecular weight distribution (MWD) is defined by a logarithmic scale on the horizontal *M* axis, and a linear scale on the vertical *w* axis. Titles on the axis are marked by different habits. On standard non-logarithmic scale as Fig. 7.4 is used for vertical axis d*W*(*M*)/d(log *M*). Also marking *w*(*M*) is used, when marks on the horizontal axis are used exponential form as 10^3, 10^4. Many times are not shown absolute values on the vertical scale as for MWD level has no meaning on the contrary the case is for normalized REDs, where values are essential.

In charts MWD is marked d*W*(*M*)/d log *M*, d*W*/d log *M* or simply *w*(log *M*). We use mainly d*W*(*M*)/d(log *M*) marking, if MWD is normalised. Different habits can be seen with charts produced by commercial GPC/SEC devices.

Normalization of RED and MWD distributions is explained in the Chap. 3.

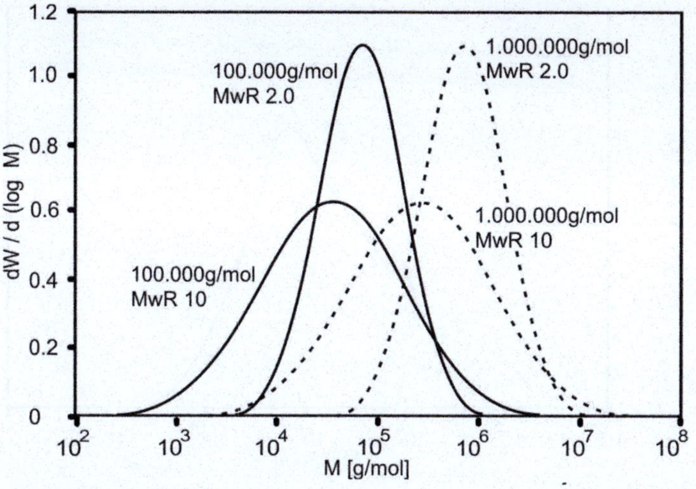

Fig. 7.4 Weight distribution of polymers with $Mw = 10^5$ g/mol and dashed line $Mw = 10^6$ g/mol by Wesslau distribution (Eq. 3.2). Used polydispersity values are $MwR = 2.0$ and $MwR = 10$

7.7 Other MWD Models

The *Zimm-Schultz distribution* model is used with polymers produced via condensation reactions or with metallocene catalysts. *Beasley distribution* and other related to the polymerization type is discussed by Graessley [4].

Bimodal and mixtures of polymers need unique data as shown with PS example at Sect. 4.2 and Malkin two-box test explained Sect. 13.5.

Figure 7.5 shows MWD computed from complex viscosity data year 2005 and used in this book with several year versions informed by GPC/SEC data.

7.8 MWD Function and Array on Spreadsheet

We present an MWD view on a spreadsheet, and the example used is 1840H LDPE polymer obtained in 2001 from BASF. Figure 7.5 shows the MWD chart developed in Excel, which was later redrawn for printing purposes. Figure 7.6 displays the top of the spreadsheet for Fig. 7.5, and the formulas are shown in Fig. 7.7.

In our examples number of used variables (index) on column cells is 200. On a top row is name of variable or distribution. The first variable is set on the row 10 as earlier.

Logarithmic rate variable x is on 0.1 steps from log(−10) to log(+10) or on columns −10, −9.9, −9.8….. done by simple addition −10 + 0.1 formula to the cells downwards.

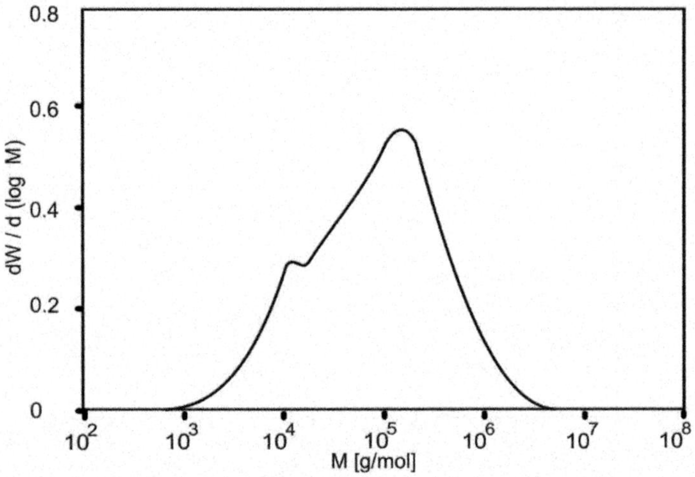

Fig. 7.5 MWD computed from complex viscosity data with year 2005 information, on spreadsheet data curve is gained Mw 226595 g/mol and MwR 14.8. Please obtain the "signal" bump on the left side for MWD inducted by derivation of complex viscosity data and left for checking if chart is MWD or RED

	B	C	D	E	F
5			MWD import		
6					
7					
8		Import data		On the standard array	
9	row	log M	w(log M)	log M	w(log M)
10	0	0.02	3.1237E-16	0	3.12367E-16
11	1	0.04	0.00003	0.05	0.001732936

Fig. 7.6 Shows measured or obtained molecular weight (*M*) data in Column A and the converted logarithmic molecular weight log M in Column B. The obtained fraction w is listed in Column C. Column D contains a standardized weight scale with 200 record items used in these examples. Column E organizes fractions using the XLOOKUP function found in Excel spreadsheets. Please note that the imported data comprises 100 items on this example

	B	C	D	E	F
5			MWD import		
6					
7					
8		Import data		On the standard	array
9	row	log M	w(log M)	log M	w(log M)
10	0	0.02	3.12366780660841E-16	0	=XLOOKUP(E10;C$10:C$110;D$10:D$110;0;1;1)
11	=+B10+1	0.04	0.00003	=+E10+0.05	=XLOOKUP(E11;C$10:C$110;D$10:D$110;0;1;1)

Fig. 7.7 Same as with Fig. 7.6, but in the Formula Auditing Mode to see used formulas. As this data imported example has 100 items or fractions, in the XLOOKUP function has imported array size B\$10:B\$110. Output array has 200 items or rows used here in examples. RED data can be imported using a similar procedure

Respectively logarithmic molar mass M is from log(0) to log(+10) or on columns −10, −9.95, −9.9….. done by simple addition −10 + 0.05.

When you make first spreadsheet, you can copy formulas to new base and add of fix only some columns.

7.9 Importing MWD Data on Spreadsheet

You may obtain MWD data in digital form, either as text or in table format. Although there are several methods to acquire MWD data, discussing them falls outside the scope of this topic. The $w(M)$ must be placed in the logarithmic molecular weight M column and the fraction w as illustrated in Fig. 7.6.

Figure 7.7 displays the top of the spreadsheet for the formulas. The imported data can vary in steps, but the row or record number must be set for the XLOOKUP function. Alternatively, the VLOOKUP function can be used, and nearly every software offers similar functions to organize data (see Fig. 7.8).

MWD data must be arranged in ascending order by utilizing the Editing/Sort & Filter command before using XLOOKUP function. The number of records may vary; for example, our sample spreadsheets here contain 200 rows. In Column D, we input our variable data, log M, which is used in the all examples.

7.9.1 List of Formulas in the Cells

Input values:

B10	=0
Column C	log M values
Column D	$w(\log M)$ fractions
E10	=0.

Functions:

E11 =+D10 + 0.05 (log M scale)

D11 =XLOOKUP(D10;B$10:B$**110**;C$10:C$**110**;0;1;1) Please obtain, the number 110 varies depending on number of fractions in imported MWD data.

7.10 Normalization, Polydispersity and Mean Value of MWD

After importing MWD to 200 arrays in our examples we normalize and calculate polydispersity and mean value of MWD to get dW/log M.

	B	C	D	E	F	G	H	I	J	K
3		MWD					Mn	15315.1		
4							Mw	226595		
5							MwR	14.8		
6										
7										
8			SUM:	7.24278	0.36214	20	0.00131	4531908	OUT MWD	
9	row	M	log M	w(log M)	w(log M)	norm dW	w/M	w*M	log M	norm dW/d log M
10	0	1.00	0.00	0	0	0	0	0	0	0
11	1	1.12	0.05	0	0	0	0	0	0	0

Fig. 7.8 Top of spreadsheet for Fig. 7.5, where *Mw* and *MwR* are computed obtained complex viscosity measurements and computation of MWD year 2005

	B	C	D	E	F	G	H	I	J	K
3		MWD					Mn	=G8/H8		
4							Mw	=+I8/G8		
5							MwR	=+I4/I3		
6										
7										
8			SUM:	=+SUM(E10:E210)	=+SUM(F10:F210)	=+SUM(G10:G210)	=+SUM(H10:H210)	=+SUM(I10:I210)	OUT MWD	
9	row	M	log M	w(log M)	w(log M)*dM	norm dW/d log M	w/M	w*M	=+D9	norm dW/d log M
10	0	=10^D10	0	0	=+E10	=+E10/F$8	=+H11	=+I11	=+D10	=+AVERAGE(G9:G11)
11	=+B10+1	=10^D11	=+D10+0.05	0	=+E11*(D11-D10)	=+E11/F$8	=+G11/C11	=+G11*C11	=+D11	=+AVERAGE(G10:G12)

Fig. 7.9 Similar to Fig. 7.8, but shown in Formula Auditing Mode to display the formulas used

Formulas are shown in Fig. 7.9, with explanations of the formulas in the cells provided in Sect. 7.9.1. The sum of each row is displayed in row 8 for reference, although not all values are necessary for the calculations.

7.10.1 List of Formulas in the Cells

Input values:

E10:E210 $w(\log M)$ distribution.

Functions:

I3 =G7/H7
I4 =I7/G7
I5 =I4/I3
E8 =SUM(E10:E210)
F8 =SUM(F10:F210)
G8 =SUM(G10:G210)
H8 =SUM(H10:H210)

I8 =SUM(I10:I210)
F10 =E10
H10 =H11
I10 =I11
C11 =10^D11
D11 =+D10 + 0.05
E11 =(fraction of imported MWD distribution)
F11 =+E11*(D11-D10)
G11 =+E11/F$8
H11 =+G11/C11
I11 =+G11*C11
J11 =+D11
K11 =+AVERAGE(G10:G12).

References

1. Borg, T., & Pääkkönen, E. J. (2009). Linear viscoelastic models. Part I. Relaxation modulus and melt calibration. *Journal of Non-Newtonian Fluid Mechanics*, *156*, 121–128.
2. Gumbel, E. J. (1958). *Statistics of extremes*. Columbia University Press.
3. Grubisic, Z., Rempp, P., & Benoit, H. (1967). A universal calibration for gel permeation chromatography. *Polym. Lett.*, *5*, 753–759.
4. Graessley, W. W. (2003). *Polymeric liquids & networks structure and properties*. Garland Science.

Chapter 8
From MWD to Red and Back Again

Back-and-forth conversions between RED and MWD distributions are needed to obtain the best *Mf* and *Hf* factors, and this procedure is utilized for many other purposes as well. Another important aspect is testing whether the conversion formulas are functioning correctly.

8.1 Used Polymer with New GPC/SEC Relations

We utilize steady shear viscosity and fresh MWD data from Poh et. al. [1] from the year 2022 for 1840H LDPE from BASF. The average molecular weight $\overline{M}_W$ = 91955 g/mol, was measured by GPC/SEC and $MwR = 4.53$. Measured MWD by GPC/SEC has changed during years as discussed in Sect. 3.9, but may have different MWDs during years, the RED is always constant related to the viscoelastic properties.

Then, we test the linearity of conversion by transforming RED back to MWD to check if the result is consistent. However, initially, we must import data suitable for our spreadsheet.

8.2 Generating Linear Relation Between MWD and RED Data

We use a single spreadsheet for developing relation between MWD and RED by finding *Mf* and *Hf* values or *Mw* and *MwR* values for MWD. In the spreadsheet example we start from MWD. Several methods exist for manually copying MWD data onto a chart. Importing MWD data on spreadsheet is discussed in Sect. 7.9. Another approach involves initially generating a Wesslau or similar distribution that matches

T. Borg, *Control Theory in Rheology*, https://doi.org/10.1007/978-3-031-88248-7_8

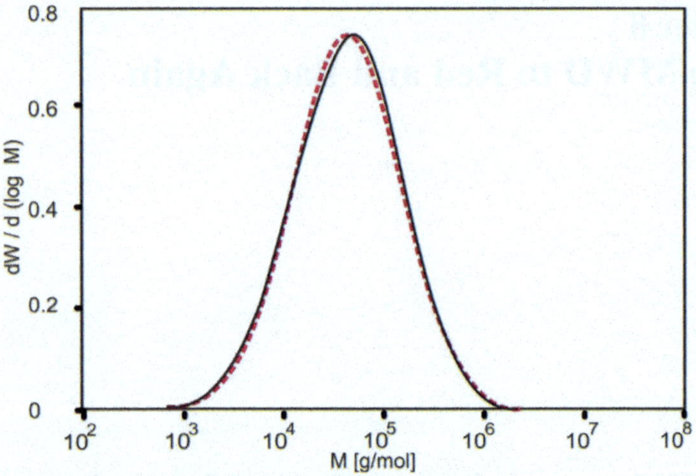

Fig. 8.1 Presents measured MWD data for LDPE 1840H from 2022, as reported by Poh et al. [1]. This data was manually inputted following the procedure outlined in Fig. 8.2 marked smooth line. A Wesslau distribution was then generated to match the same average molecular weight and polydispersity index marked dashed line

the average molecular weight and polydispersity value. Subsequently, the curve's shape can be modified and smoothed using mean functions such as AVERAGE, following the procedures discussed in Chap. 3.

In the spreadsheet examples, the Wesslau distribution is utilized because it can be generated using an equation. This method provides a systematic and calculable approach to creating distributions that can be easily manipulated and analysed within the spreadsheet environment, facilitating the examination and comparison of MWD and RED distributions.

Earlier were shown in Sect. 3.10 Spreadsheet example for generating RED by Wesslau distribution, but now is presented some add cells.

Figure 8.2 shows top of spreadsheet for creating MWD distribution as close as possible with measured GPC/SEC 1840H LDPE shown in Fig. 8.1. Figure 8.3 shows same cells in the Formula Auditing Mode.

8.2.1 List of Formulas in the Cells

Input values:

C5 4.635 (μ is the mean)
C6 0.534 (σ is standard deviation).

Functions:

	B	C	D	E	F	G	H	I	J	K
3		MWD					Mn	20263		
4							Mw	91896		
5	myy	4.64					MwR	4.54		
6	sign	0.53								
7										
8			SUM:	20	1	20	0.00099	1837926	OUT MWD	
9	row	M	log M	w(log M)	w(log M)*dM	norm dW/d log M	w/M	w*M	log M	norm dW/d log M
10	0	1.00	0	3.2645E-17	3.26454E-17	3.26454E-17	6.5E-17	8.2E-17	0	5.29533E-17
11	1	1.12	0.05	7.3261E-17	3.66307E-18	7.32613E-17	6.5E-17	8.2E-17	0	8.96271E-17

Fig. 8.2 By Wesslau distribution is generated MWD shown in Fig. 8.1. Used $\mu = 4.635$ and $\sigma = 0.534$. Calculated average molecular weight $Mw = 91{,}896$ g/mol and $MwR = 4.54$ close with GPC/SEC measurements

	B	C	D	E	F	G	H	I	J	K
3		MWD					Mn	=G8/H8		
4							Mw	=+I8/G8		
5							MwR	=+I4/I3		
6	myy	4.635								
7	sigma	0.534								
8			SUM:	=+SUM(E10:E210)	=+SUM(F10:F210)	=+SUM(G10:G210)	=+SUM(H10:H210)	=+SUM(I10:I210)	OUT MWD	
9	row	M	log M	w(log M)	w(log M)*dM	norm dW/d log M	w/M	w*M	=+D9	norm dW/d log M
10	0	=10^D10	0	=1/($C$7*SQRT(2*PI()))*EXP(-1/2*((LOG10(C10)-$C$6)/$C$7)^2)	=+E10	=+E10/F$8	=+H11	=+I11	=+D10	=+AVERAGE(G9:G11)
11	=+B10+1	=10^D11	=+D10+0.05	=1/($C$7*SQRT(2*PI()))*EXP(-1/2*((LOG10(C11)-$C$6)/$C$7)^2)	=+E11*(D11-D10)	=+E11/F$8	=+G11/C11	=+G11*C11	=+D11	=+AVERAGE(G10:G12)

Fig. 8.3 Displays the same content as Fig. 8.2, but in Formula Auditing Mode to showcase the formulas used

```
I3   =G8/H8
I4   =I8/G8
I5   =I4/I3
E8   =SUM(E10:E210)
F8   =SUM(F10:F210)
G8   =SUM(G10:G210)
H8   =SUM(H10:H210)
I8   =SUM(I10:I210)
F10  =E10
H10  =H11
I10  =I11
C11  =10^D11
D11  =+D10 + 0.05
E11  =1/($C$6*SQRT(2*PI()))*EXP(−1/2*((LOG10(C11) − $C$5)/$C$6)^2)
F11  =+E11*(D11-D10)
G11  =+E11/F$8
H11  =+G11/C11
I11  =+G11*C11
J11  =+D11
K11  =+AVERAGE(G10:G12).
```

8.3 RED from MWD Data on Spreadsheet

Next task is to convert the MWD into the RED function using the melt calibration principle according to Sect. 7.1 as demonstrated in the spreadsheet in Fig. 8.4, with the corresponding formulas shown in Fig. 8.5.

Fig. 8.5Displays the same content as Fig. 8.4, but in Formula Auditing Mode to showcase the formulas used

8.3.1 List of Formulas in the Cells

Input values:

C5 1.63 (*Hf*)
C6 6500 (*Mf*)
M10 10.

Functions:

S8 =SUM(S10:S210)
T8 =SUM(T10:T210)

	M	N	O	P	Q	R	S	T	U	V	W
3		RED									
4											
5	*hf*	1.63									
6	*Mf*	6500	g/mol								
7											
8						SUM:	20	16.29	1.63	OUT	10
9	row	log x	M	log M	x	log x	f(log x)	tilde w(log x)	f(log x)*x	log x	norm dW/d log x
10	0	10	1.00	0.00	1640774	6.22	5.3E-17	0	0	10	0
11	1	9.9	1.12	0.05	1360031	6.13	9E-17	0	0	9.9	0

Fig. 8.4 By Melt Calibration principle with *Hf* = 1.63 and *Mf* = 6500 g/mol is converted MWD to RED. Polymer structural *Mf* sets the average molecular weight$\overline{M}_W$ and conversion factor *Hf* between rate and molar weight scales set *MwR* called polydispersity value

	M	N	O	P	Q	R	S	T	U	V	W
3		RED									
4											
5	*hf*	1.63									
6	*Mf*	6500	g/mol								
7											
8						SUM:	=+SUM(S10:S210)	=+SUM(T10:T210)	=+SUM(U10:U210)	OUT RED	=+SUM(W10:W210)
9	row	log x	M	log M	x	log x	f(log x)	tilde w(log x)	f(log x)*x	=+N9	norm dW/d log x
10	0	10	=10^P10	0	=+(N$6/O10)^N$5	=+LOG(Q10)	=+K10	=XLOOKUP(N10;R$10:R$210;S$10:S$210;0;1;1)	=+T10	=+N10	=+AVERAGE(T9:T11)/U$8
11	=+M10+1	=+N10-0.1	=10^P11	=+P10+0.05	=+(N$6/O11)^N$5	=+LOG(Q11)	=+K11	=XLOOKUP(N11;R$10:R$210;S$10:S$210;0;1;1)	=+T11*(N10-N11)	=+N11	=+AVERAGE(T10:T12)/U$8

Fig. 8.5 Displays the same content as Fig. 8.4, but in Formula Auditing Mode to showcase the formulas used

U8 =SUM(U10:U210)
M11 =+M10 + 1
N11 =+N10 – 0.1
O11 =10^P11
Q11 =+(N$6/O11)^N$5 Melt Calibration formula
R11 =+LOG(Q11)
S11 =+K11 Importing MWD
T11 =XLOOKUP(N11;R$10:R$210;S$10:S$210;0;1;1)
U11 =+T11*(N10 – N11)
V11 =+N11 To get abscissa for the chart
V11 =+AVERAGE(T10:T12)/U$8 You can wider average field.

8.4 MWD from RED Data on Spreadsheet

On the right direction of the spreadsheet, we convert MWD back from RED to check if formulas are done right. We get Fig. 8.8 with data on spreadsheet as follows (see Fig. 8.7).

8.4.1 List of Formulas in the Cells

Input values:

AA5 1.63 (*Hf*)

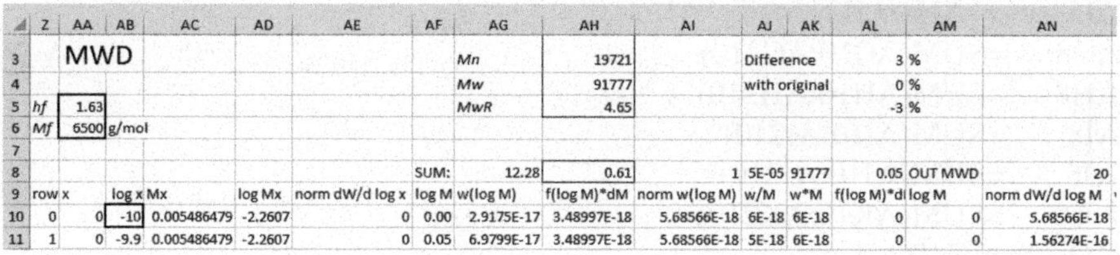

	Z	AA	AB	AC	AD	AE	AF	AG	AH	AI	AJ	AK	AL	AM	AN
3		MWD						Mn	19721		Difference		3	%	
4								Mw	91777		with original		0	%	
5	hf	1.63						MwR	4.65				-3	%	
6	Mf	6500	g/mol												
7															
8							SUM:	12.28	0.61	1	5E-05	91777	0.05	OUT MWD	20
9	row	x	log x	Mx	log Mx	norm dW/d log x	log M	w(log M)	f(log M)*dM	norm w(log M)	w/M	w*M	f(log M)*dl	log M	norm dW/d log M
10	0	0	-10	0.005486479	-2.2607	0	0.00	2.9175E-17	3.48997E-18	5.68566E-18	6E-18	6E-18	0	0	5.68566E-18
11	1	0	-9.9	0.005486479	-2.2607	0	0.05	6.9799E-17	3.48997E-18	5.68566E-18	5E-18	6E-18	0	0	1.56274E-16

Fig. 8.6 MWD converted back from RED data of 1840H viscosity measurements. The error with original MWD Fig. 8.4 for *Mw* is 0%, *Mn* 3% and *MwR* −3%. With larger arrays or more rows the error should be less

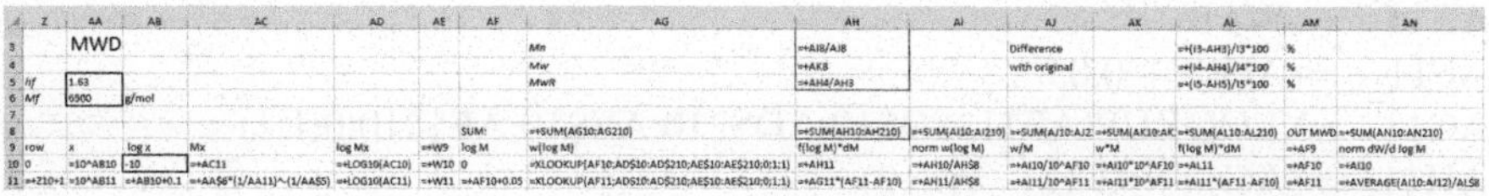

	Z	AA	AB	AC	AD	AE	AF	AG	AH	AI	AJ	AK	AL	AM	AN
3		MWD						Mn	=+AI8/AJ8		Difference		=+(I3-AH3)/I3*100	%	
4								Mw	=+AK8		with original		=+(I4-AH4)/I4*100	%	
5	hf	1.63						MwR	=+AH4/AH3				=+(I5-AH5)/I5*100	%	
6	Mf	6500	g/mol												
7															
8							SUM:	=+SUM(AG10:AG210)	=+SUM(AH10:AH210)	=+SUM(AI10:AI210)	=+SUM(AJ10:AJ2	=+SUM(AK10:AK	=+SUM(AL10:AL210)	OUT MWD	=+SUM(AN10:AN210)
9	row	x	log x	Mx	log Mx	=+W9	log M	w(log M)	f(log M)*dM	norm w(log M)	w/M	w*M	f(log M)*dM	=+AF9	norm dW/d log M
10	0	=10^AB10	-10	=+AC11	=+LOG10(AC10)	=+W10	0	=XLOOKUP(AF10;AD$10:AD$210;AE$10:AE$210;0;1;1)	=+AH11	=+AH10/AH$8	=+AI10/10^AF10	=+AI10*10^AF10	=+AL11	=+AF10	=+AI10
11	=+Z10+1	=10^AB11	=+AB10+0.1	=+AA$6*(1/AA11)^(1/AA$5)	=+LOG10(AC11)	=+W11	=+AF10+0.05	=XLOOKUP(AF11;AD$10:AD$210;AE$10:AE$210;0;1;1)	=+AG11*(AF11-AF10)	=+AH11/AH$8	=+AI11/10^AF11	=+AI11*10^AF11	=+AI11*(AF11-AF10)	=+AF11	=+AVERAGE(AI10:AI12)/AL$8

Fig. 8.7 Displays the same content as Fig. 8.6, but in Formula Auditing Mode to showcase the formulas used

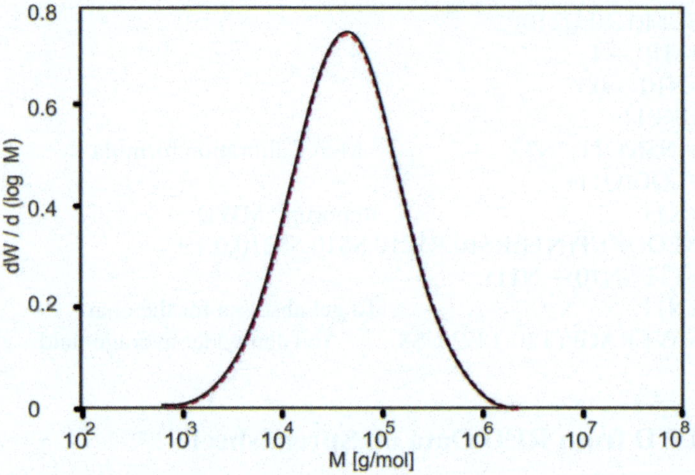

Fig. 8.8 Presents the original MWD generated by the Wesslau formula, converted to RED, and then back to a new MWD, indicated by dashed lines. The computation was performed using the spreadsheets discussed in this section. The error between the original and final MWD distribution after conversions was a minimal 0.17%

AA6 6500 (*Mf*)
Z10 0
Ab10 −10

Functions:

AI3	=+AI8/AI8	
AI4	=+AK8	
AI5	=+AH4/AH3	
AG8	=SUM(AG10:AG210)	
AH8	=SUM(AH10:AH210)	
AI8	=SUM(AI10:AI210)	
AJ8	=SUM(AJ10:AJ210)	
AK8	=SUM(AK10:AK210)	
Z11	=+Z10 + 1	
AA11	=10^AB11	
AB11	=+AB10 + 0.1	
AC11	=+AA$6*(1/AA11)^−(1/AA$5)	Melt Calibration formula
AD11	=+LOG10(AC11)	
AE11	=+W11	
AF11	=+AF10 + 0.05	
AG11	=XLOOKUP(AF11;AD$10:AD$210;AE$10:AE$210;0;1;1)	
AH11	=+AG11*(AF11 − AF10)	
AI11	=+AH11/AH$8	
AJ11	=+AI11/10^AF11	

Table 8.1 Variables: mean (μ) and variance (σ)

	μ	σ	Mf	Hf
Shear viscosity	−1.4	0.9	7000	1.7
Complex viscosity	−0.9	1.92	35000	3.6
Relaxation modulus	−0.9	1.92	35000	3.6

```
AK11  =+AI11*10^AF11
AL11  =+AI11*(AF11 – AF10)
AM11  =+AF11                     To get abscissa for the chart
AL11  +AVERAGE(AI10:AI12)/AL$8   You can wider average field
AN11  =+K11                      For comparison with the original MWD.
```

8.5 Error Generated by MWD-RED-MWD Conversion

Figure 8.8 shows the computation results after converting the original MWD to RED and back to MWD using the spreadsheets presented in this chapter. The original MWD is generated using the Wesslau formula.

The error between the original MWD and the MWD-RED-MWD conversions is negligible, even with a small sample of 200 items, and is primarily due to rounding errors. This demonstrates that the conversion procedure is linear.

8.6 Used Conversion Factor for Different Flows

In Table 8.1 is shown variables mean μ and variance σ to generate Wesslau distribution for RED. Another route is to use polymer structural *Mf* and polydispersity *Hf* values, which are depending on the chemical nature of polymer.

Variables: Mean μ and variance σ to for generating Wesslau distribution and generating RED by Melt Calibration from MWD with *Mf* and *Hf* values for LDPE.

Reference

1. Poh, L., Wu, Q., Chen, Y., & Narimissa, E. (2022). Characterization of industrial low-density polyethylene: A thermal, dynamic mechanical, and rheological investigation. *Rheologica Acta, 61*, 701–720.

Chapter 9
From Measured Viscosity Data to MWD

We have successfully modelled various viscoelastic flows starting from the Rheologically Effective Distribution (RED), and now, thanks to the linear Control Theory (CT) and *Unified* formulas, we can explore new directions to obtain Molecular Weight Distribution (MWD).

We've found that measured complex viscosity alone is sufficiently accurate for a while to extract both RED and MWD, although, in principle, shear viscosity and relaxation modulus data are also usable. However, it's important to note that using the Analytical formula type, such as Eq. (9.6), may be challenging due to its nature as an ill-posed problem. To navigate this challenge and compute MWD, we relied on Unified Eqs. (9.2), (9.4), and (9.5) for different flows. Despite these complexities, we provide a historical background and present the Analytical formula (9.6) for reference. This chapter draws exclusively on published articles. [1, 2]. Given the narrow data range required for simple shear viscosity, we utilize it in the spreadsheet example.

9.1 RED $w(\omega)$ from Complex Viscosity Measurements

We have to extract RED distribution $w(\omega)$ from model for complex viscosity $\eta^*(\omega)$ with constant characteristic frequency ω_c as

$$\log \frac{\eta^*(\omega)}{\eta_0^*} = -\log \frac{\omega}{\omega_c} \int_{-\infty}^{\log \omega} \left(P' w'(\log \psi) + P'' w'' \left(\log \frac{\psi}{R} \right) \right) d \log \psi \tag{9.1}$$

To use viscosity measurements to detect the effective distribution, $w(\omega)$, we have to solve Eq. (9.1) for RED function $w(\log \omega)$ by deriving Unified (characteristic) distribution as follows:

T. Borg, *Control Theory in Rheology*, https://doi.org/10.1007/978-3-031-88248-7_9

$$w'(\log \omega) = -\frac{d}{d \log \omega}\frac{1}{P'}\left(\frac{\log \frac{\eta^*}{\eta_0^*}}{\log \frac{\omega}{\omega_c}} + P'' \int_{-\infty}^{\log \omega} \left(w''\left(\log \frac{\psi}{R}\right)\right) d \log \psi\right) \tag{9.2}$$

Normally, distribution w(log ω) and its copy $w\left(log \frac{\omega}{R}\right)$ have zero or minimal overlap. This happens rarely with a broad MWD and a high polydispersity index (MwR > 20), but can be easily managed during the computation by performing the subtraction have $log\omega - w\left(log \frac{\omega}{R}\right)$ before conversion for the output MWD. This overlapping situation may have an effect on the computed values of P' and P''.

9.2 RED Function from Shear Viscosity Measurements

RED function $w(\log \dot{\gamma})$ is shown in this Section. Recently obtained that detecting MWD directly from shear viscosity measurements should be best alternative, if we can get relative data at different rates. The needed width of shear rate scale is about three decades to get most MWDs in shar rate ranges from 0.001/s up to 1/s or more. Derivation of shear viscosity needs relatively accurate data means, that there is not much deviation of the theoretical right values. Absolute level is not important as viscosity values are normalized by relation $\frac{\eta(\dot{\gamma})}{\eta_0}$. Also measured data can be in other physical dimensions as viscosity in Pas, but shear rate scale is important. Drawback is the fact that we do not get simultaneously wide modelled viscosity flow curve. No doubt measurements have to do in increasing steps maybe on log 0.2 steps.

Formula for detecting RED from viscosity data is shown in Eq. (9.4), where shear viscosity $\eta(\dot{\gamma})$ with constant characteristic frequency $\dot{\gamma}_c$ is modelled as follows:

$$\log \frac{\eta(\dot{\gamma})}{\eta_0} = -\log \frac{\dot{\gamma}}{\dot{\gamma}_c} \int_{-\infty}^{\log \dot{\gamma}} \left(P' w'(\log \psi) + P'' w''\left(\log \frac{\psi}{R}\right)\right) d \log \psi \tag{9.3}$$

To use viscosity measurements to detect the effective distribution, $w'(\log \dot{\gamma})$, we have to solve Eq. (9.3) for RED function $w(\log \dot{\gamma})$ by deriving Unified distribution as follows:

$$w'(\log \dot{\gamma}) = -\frac{d}{d \log \dot{\gamma}}\frac{1}{P'}\left(\frac{\log \frac{\eta}{\eta_0}}{\log \frac{\dot{\gamma}}{\dot{\gamma}_c}} + P'' \int_{-\infty}^{\log \dot{\gamma}} \left(w''\left(\log \frac{\psi}{R}\right)\right) d \log \psi\right) \tag{9.4}$$

No doubt there may be instrumental difficulties and if people gets increasing viscosity values at low rates, they believe that results are false. In the modelling shear viscosity, many times it is a little increasing before shear thinning effects, as shown in Fig. 9.4.

9.3 Effective Distribution $w(t)$ from Relaxation Modulus $G(t)$

To use relaxation modulus measurements, have fundamental problems discussed in Sect. 8.3. Anyway, we show formula to detect the effective distribution, $w(t)$, have some we have to solve similar way as above Eqs. (9.2) or (9.4) for RED function w(log t) by deriving Unified distribution $w(t)$ as follows:

$$w'(\log t) = -\frac{d}{d\log t}\frac{1}{P'}\left(\frac{\log\frac{G(t)}{G_o}}{\log\frac{\tau}{\tau_c}} + P''\int_{-\infty}^{\log\tau}\left(w''\left(\log\frac{\psi}{R}\right)\right)d\log\psi\right) \tag{9.5}$$

Normally, distribution $w(\log\,\omega)$ and its copy $w\left(log\frac{\omega}{R}\right)$ have zero or minimal overlap. This happens rarely with a broad MWD and a high polydispersity index (MwR > 20), but can be easily managed during the computation by performing the subtraction have $log\omega - w\left(log\frac{\omega}{R}\right)$ before conversion for the output MWD. This overlapping situation may have an effect on the computed values of P' and P''.

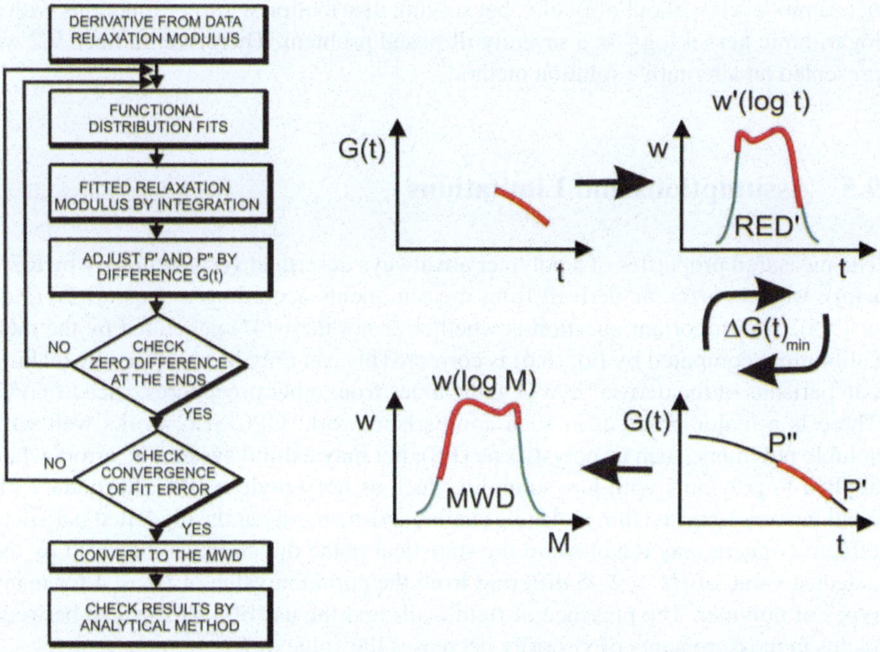

Fig. 9.1 Flowchart for the algorithm used to determine the MWD and the modelled relaxation modulus on a logarithmic scale and respective data flowchart of the procedure. Thick lines show derivative results, and thin lines are the results for the unified and analytical models. The best fit to $\Delta G(t)_{\min}$ is obtained by a standard numerical least-squares computation. Borg and Pääkkönen [1]

9.4 Analytical Formula

Only for comparison and testing MWD and RED we show Analytical formula (6.2). Bersted et al. made one of the most successful early attempts to find a relationship between the viscosity flow curve and molecular weight distribution (MWD), using the partition model [3–9]. The inverse problem of how to determine the MWD from viscosity was described by Malkin [10, 11], and is known to be an ill-posed problem, before we found above Equ. (9.2). To overcome this several authors [12–18] have used the double reptation model or the general mixing rule as analysed by Anderssen and Mead [19]. Thimm et al. [20] provided the first analytical relation for deriving the MWD from the modelled relaxation spectrum, even though this is impossible to measure directly. Analytical Eq. (9.6) for complex viscosity

$$\log \frac{\eta^*(\omega)}{\eta_0^*} = -\int_{-\infty}^{\log\omega} \left(P' w'(\log\psi) + P'' w''\left(\log\frac{\psi}{R}\right)\right) \log\frac{\omega}{\psi} d\log\psi \tag{9.6}$$

where the scalar values are $0 < P' < 40$ for entanglement and $0 < P'' < 40$ for the Rouse relaxation ranges. The convolution integral Eq. (9.6) for complex viscosity $\eta^*(\omega)$ models it without difficulty, but solving distribution $w'(\omega)$ or functions with a logarithmic kernel $\log\frac{\omega}{\psi}$ is a severely ill-posed problem. Therefore, in Sect. 9.2 we presented an alternative solution method.

9.5 Assumptions and Limitations

The measured properties of a polymer are always described very accurately by RED $w(\omega)$, $w(\dot{\gamma})$ or $w(t)$, as derived from measurements according to Eqs. (9.2), (9.4) or (9.5). The important question is whether or not the $w(M)$ generated by the melt calibration, computed by Eq. (6.6) is correct. This can only be answered by making comparisons of the derived MWD with values from other procedures, such as GPC. There is a major problem in such comparison work: GPC/SEC works well with soluble polymers, such as polystyrene (PS), but may exhibit systematic errors when applied to polymers with low solubility, such as polyethylene (PE). Secondary PE contains very long and thin molecule chains, and during shear the modelled statistical effective sphere may tend toward the statistical plane direction, as indicated by the detected value of $Hf = 2.05$ differing from the common value of $Hf = 4$ for many types of polymer. The presence of rigid walls and the use of a narrow gap between plates in measurements of viscosity decreases the value of Hf.

We can measure viscosity only over a limited rate or frequency range, but this can be extended using our models. The accuracy of the method deteriorates as the true polydispersity of the MWD increases. Master curves for viscosity generated by time–temperature superposition conversions have distortion points for derivations at junction points of different temperatures.

9.6 Computation Description

The principal algorithm used to implement the presented formulas is described on a step-by-step basis based on the measured complex viscosity data η^*_{obs} as shown in the flowchart in Fig. 9.2.

As indicated in Fig. 9.2, the computation is performed in three phases: initialization, iteration, and best-fit routines. The initialization routine involves the following steps:

1. Measured complex viscosity data ($\eta^*{}_{obs}$) is imported from the database.
2. Initial values are given for structural factor *Mf* and the relation exponent for the *Hf* scales, and random values are used for entanglement P' and the Rouse P'' factor.
3. The observed η^*_{obs} data are solved according to Eq. (9.2) by obtaining the observed characteristic effective distribution RED_{obs} $w(\log\,\omega)_{obs}$ segment. During the derivation there is a discontinuity at ω_c, which must be omitted.
4. Distribution-segment fits to $w(\log\,\omega)$ fit are approximated on both sides of the $w(\log\,\omega)_{obs}$ curve. The procedure can be performed by extrapolating from $w(\log\,\omega)_{obs}$ lines, fitting the normalized distribution with $w(\log\,\omega)_{obs}$, or using a priori information for the MWD. When the MWDs are narrow compared to the respective frequency range, segment fits do not have any effect since $w(\log\,\omega)_{fit} = 0$.
5. The RED $w(\log\,\omega)$ is normalized.
6. The Rouse distribution $w\left(log\,\frac{\omega}{R}\right)$ is generated simply by copying RED $w(\log\,\omega)$, since the polymer is the same.

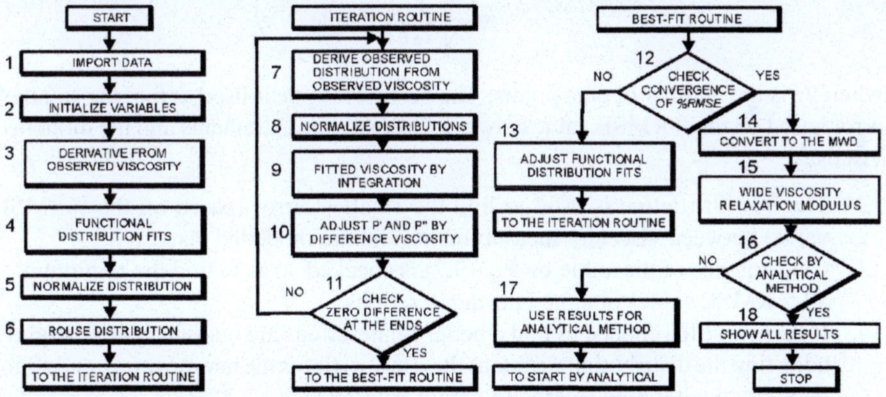

Fig. 9.2 Flowchart for the algorithm used to determine the MWD and the modelled viscoelastic properties. Borg and Pääkkönen [1]

9.7 Iteration Process

The iteration routine involves the use of several loops to accurately fit the modelled viscosity with that measured at the lowest and highest frequencies:

7. The η^*_{obs} data are solved according to Eq. (9.2) to generate $w(\log \omega)_{\text{obs}}$.
8. Distributions RED $w(\log \omega)$ and $w\left(log \frac{\omega}{R}\right)$ are normalized.
9. The wide-range fit η^*_{fit} is calculated from $w(\log \omega)$ and $w\left(log \frac{\omega}{R}\right)$ by integrating according to Eq. (9.1).
10. The values of P' and P'' are adjusted by a bracketing procedure to reduce the values of the lowest and highest frequency points of the difference viscosity curve ($\Delta\eta^* = \eta^*_{obs} - \eta^*_{fit}$) to zero within an assigned error.
11. This Iteration Loop is Performed Until the Highest and Lowest Data Points Are Equal to the Modelled Viscosity Values, or Numerically $\Delta\eta^* = 0$ at Those Points.

9.8 Fitting Routines

Least-square procedures are widely used in numerical computations. Here, the quality of the fit between the observed η_{obs} and predicted η_{fit} viscosity values was quantified using a modified least-square procedure called the percentage root-mean-square error function (%RMSE) [21]:

$$\%RMSE = \frac{100}{N}\sqrt{\sum_{i=1}^{N}\left(\frac{\eta^*_{obs} - \eta^*_{fit}}{\eta^*_{obs}}\right)^2} \tag{9.7}$$

where N is the number of data points. The best-fit routine is used to find the smallest error based on the %RMSE value between viscosity measurements and the modelled viscosity.

12. The best-fit routine is used to find the smallest error (based on the %RMSE value) between viscosity measurements and the modelled viscosity.
13. Convergence of the value by Eq. (9.7) is checked so as to find the minimum of all %RMSE values obtained during iterations.
14. If there is still a chance to find a better fit, iterations are continued after slightly adjusting the distribution-segment fit $w(\log \omega)$ fit for the new position, and again applying the iteration routine to obtain the RED.
15. When the best fit is found, we convert $w(\log \omega)$ to the final MWD $w(\log M)$ using Mf and Hf values by the melt-calibration concept according to Eq. (9.7).
16. Check if the Analytical method described by Eq. (9.4) has already been used and if needed. This step may be unnecessary.

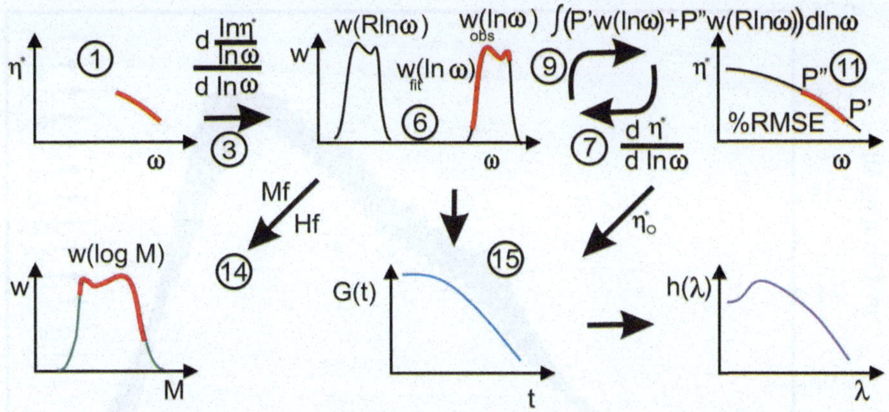

Fig. 9.3 Data flowchart of the procedure. Direct results are shown by thicker lines on log scales, except for the w scale, and the thinner lines are the modelled values. Numbers inside circles correspond to the steps in Fig. 9.1. Borg and Pääkkönen [1]

17. Convert constants used in the Analytical method into a new complete loop, using the same procedure as for the Unified method as much as possible. This improves the likelihood of solving Analytical Eq. (9.4) to obtain the RED and MWD with standard solution methods based on a priori information using parallel computing. This step may be unnecessary.
18. Show and Compare Results Obtained from Both the Fast Unified (Eq. 9.2) and Analytical (Eq. 9.4) Methods. This Step May Be Unnecessary.

Also, wide-range and accurate viscosity fits and relaxation modulus $G(t)$ are computed according to Eq. (6.11), and spectrum $h(\lambda)$ is computed according to Eq. (6.15). Figure 9.3 shows the results.

The structural and conversion factors (Mf and Hf) between scales are found for new polymer types by comparing the results obtained with GPC and SEC. Mf and Hf influence only the average molecular weight $\overline{M}w$ and MwR, respectively. A segment of the distribution related to the MWD is gained directly by derivation from the measured data, and then several rheological properties are computed by integrating the derived and modelled distributions.

The formulas contain a variable in the exponent, and hence in the numerical computation practice we need to consider logarithmic and exponential functions simultaneously during cyclic derivation, integration, and for other computations – this power-integral-logarithm-derivation processing is a very unstable technique due to the presence of some difference between the measurements and the model values. We therefore adapted the standard numerical methods found in textbooks for this case.

The MWDs of LDPE computed at 13 temperatures are summarized in Fig. 9.4. This test revealed that temperature had a minor effect on the values of the constants:

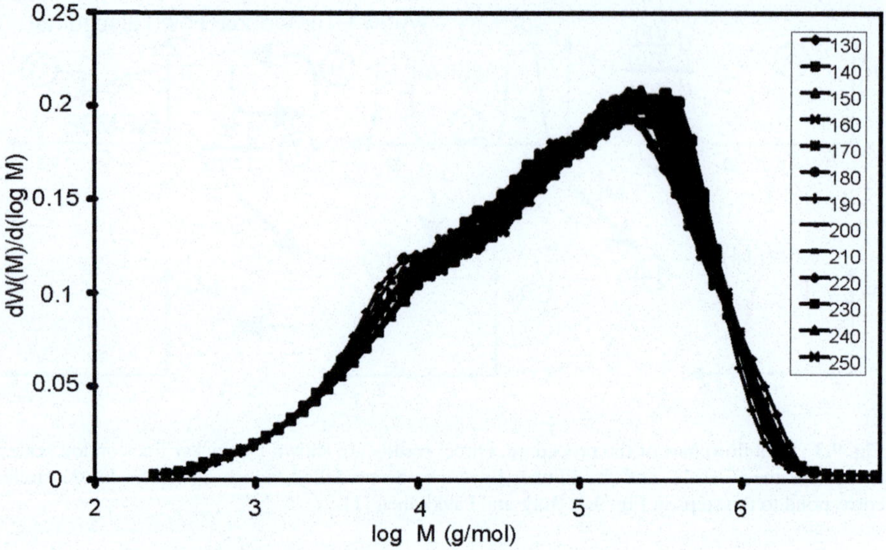

Fig. 9.4 Summary of Lupolen 1840H LDPE MWDs computed from data measured at temperatures from 130 to 250 °C (in 10 °C increments). The deviation in $\overline{M}w$ was 2.3%, indicating that the procedure was repeatable. Borg and Pääkkönen [1]

$$Mf = Mf_0 + (T - T_0)250 \tag{9.8}$$

The average obtained $\overline{M}$w by computations was 234,689 (with a deviation of 2.3%) and MwR = 14.1 for all computations.

Respective measured and fitted wide viscosity curves are shown in the Sect. 5.4 and Fig. 9.4 as a result of several iterations.

In the Fig. 9.5 is shown respective measured and modelled viscosity flow curves as a result during thousands of iterations.

The error between modelled and measured data is discussed in the Sect. 13.7.

9.9 An Example of Obtaining RED from Shear Viscosity Data

We derive RED from steady shear viscosity data, though it is semi-measured and composed from different sources, as discussed in Chapter 4. The spreadsheet addition presented here complements the example introduced in Sect. 4.13, "Example on a Spreadsheet." In fact, the modelling of the flow curve and the computation of RED must be performed simultaneously, as mentioned earlier. This same addition can also be applied to the modelling of complex viscosity spreadsheet introduced in

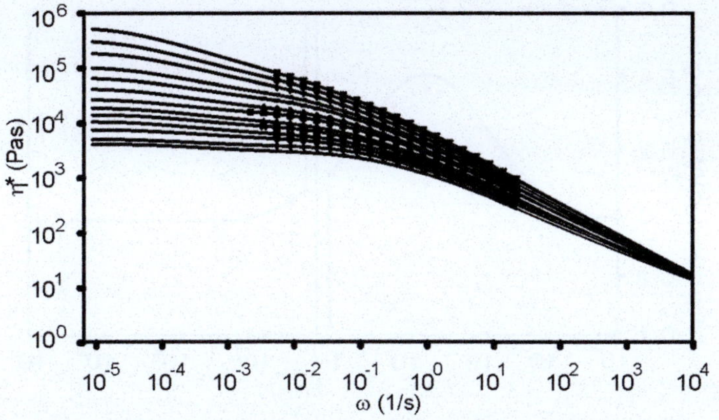

Fig. 9.5 Measured viscosity curves for Lupolen 1840H LDPE at temperatures ranging from 130 to 250 °C in 10 °C increments. The points represent measured data provided by BASF, and the curves shown over a wider scale were modelled using software. The average error (%RMSE) between the measured and modelled viscosities is 0.0009%. the same figure was shown in the Sect. 5.4 as a result of several iterations. Borg and Pääkkönen [1]

Sect. 5.7 and by Eq. 9.2, as well as to the relaxation modulus spreadsheet in Sect. 6. 8 and by Eq. 6.12.

In the presentation of the computation procedure [2], complex viscosity was used, as this data is more readily available. However, obtaining RED and subsequently MWD from shear viscosity is preferable, as it scans a wider RED distribution at the same rate range, for reasons discussed elsewhere in this text.

We formulate Eq. (9.9) for the RED function $w(\log \dot{\gamma})$ by deriving the Unified (characteristic) distribution from steady shear viscosity as follows:

$$w'(\log \dot{\gamma}) = -\frac{d}{d \log \dot{\gamma}} \frac{1}{P'} \left(\frac{\log \frac{\eta}{\eta_0}}{\log \frac{\dot{\gamma}}{\dot{\gamma}_c}} + P'' \int_{-\infty}^{\log \dot{\gamma}} \left(w'' \left(\log \frac{\psi}{R} \right) \right) d \log \psi \right) \tag{9.9}$$

9.9.1 On a Spreadsheet

We use same spreadsheet and data as for Sect. 4.13 to model steady shear viscosity measurements of 1840H LDPE using artificial Wesslau distribution for REDs. For Wesslau distribution is used same values $\mu = -1.4$ and $\sigma = 0.9$. Additionally, the elastic value $P' = 0.40$ and viscous $P'' = 0.22$ were maintained consistently.

Please obtain that during true computation by derivation of viscosity curve is not only achieved RED, but P', P'' and R is obtained during fitting process as described earlier. Results are shown in the Fig. 9.6.

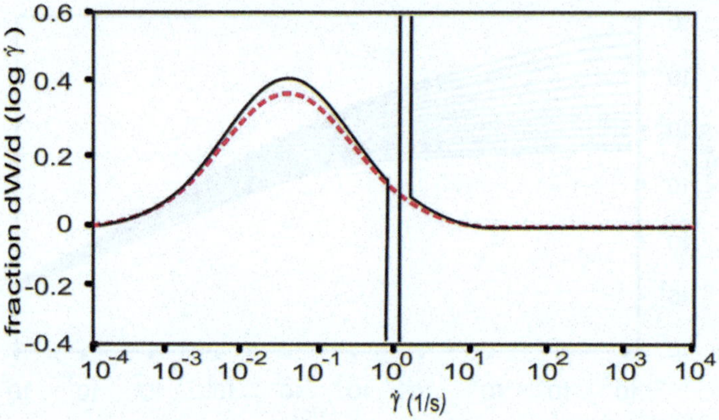

Fig. 9.6 On the MS Excel 365 spreadsheet cut and redrawn chart as it is derivation of viscosity curve. Original RED is also included shown by dashed line. During the derivation there is a discontinuity at $\dot{\gamma}_C$, which must be omitted and distribution normalised

9.9.2 *Formulas on the Spreadsheet*

Attached Fig. 9.7 is cut of pre-prepared calculation spreadsheet with formulas of the same computation as get Fig. 9.5. In the Sect. 3.10 and earlier chapters is explained some general principles for presented spreadsheets. Figure 9.8 shows same cells in the Formula Auditing Mode.

	Q	R	S	T	U
3	RED				
4					
5	P'	0.4		x/0	1
6	P''	0.22			
7	log eta/o	4.05			
8					
9	η/fit	log x	preRED	RED/fit	RED/orig
10	4.05	**-10.0**	-1.854E-16	0	2.6E-21
11	4.05	-9.9	-1.854E-16	0	7.6E-21
12	4.05	-9.8	-8.837E-17	-9.7E-16	2.2E-20

Fig. 9.7 Presents a top Section of an MS Excel 365 spreadsheet used for deriving RED from viscosity flow curve. Cells outlined with thick borders and containing bold numbers are designated for input data. The RheoPower software internally determines these values when computing viscosity measurements data. Cells framed by thinner borders represent output data or are intended solely for verification purposes

	Q	R	S	T	U
3	RED				
4					
5	=+H5	=+I5		=+K5	=+L5
6	=+H6	=+I6			
7	=+H7	=+I7			
8					
9	η/fit	=+D9	preRED	RED/fit	RED/orig
10	=+Q11	**=+D10**	=+S11	=+T11	=+H10
11	=+LOG(POWER(10;I$7)*POWER(C11;-P11))	=+D11	=(1/R$5*(LOG((10^Q11)/(10^R$7))/LOG((10^R11)/U$5)+O11))	=-(S10-S11)/(R10-R11)	=+H11
12	=+LOG(POWER(10;I$7)*POWER(C12;-P12))	=+D12	=(1/R$5*(LOG((10^Q12)/(10^R$7))/LOG((10^R12)/U$5)+O12))	=-(S11-S12)/(R11-R12)	=+H12

Fig. 9.8 Formulas to get RED. Cells outlined with thick borders and containing bold numbers are designated for input data. Cells framed by thinner borders represent output data or are intended solely for verification purposes

Input values:

R5 0.4 (P' elastic value).
U5 1 ($\dot{\gamma}_c$).
R6 0.22 (P'' viscosity value).
R7 4.05 (Zero viscosity η_0 at shear rate 1/s).
R10 −10 (initial log rate for rows and arrays).

Formulas:

Q11 = + LOG(POWER(10;I$7)*POWER(C11;-P11)).
R11 = + D11.
S11 = (1/R$5*(LOG((10^Q11)/(10^R$7))/LOG((10^R11)/U$5) + O11)).
T11 = −(S10-S11)/(R10-R11).

Option input and formulas:

U11 = + H11.

9.9.3 Computing RED from Viscoelastic Measurements

In the spreadsheet example, calculating RED appears straightforward compared to modelling flows. We utilized a viscosity curve modelled from RED, but in practice, computing from practical measurements, even tiny errors in data can lead to significant errors in the RED form. Absolute errors in viscosity levels are inconsequential, but relative errors arising from derivation are significant.

The discontinuity point poses the least concern; however, software codes have been developed to offer different solutions for mitigating errors. Caution must be exercised when smoothing RED.

In principle, shear viscosity data are more sensitive and may lead to more fluctuations in RED compared to complex viscosity analyses.

In summary, to complete the palette and procedure, everything must be well-founded and interrelated through several iterations to achieve the best fit for the complete computation. The derivation presented in Sect. 9.9.2 alone suggests an unrealistic ease of work.

References

1. Borg, T., Pääkkönen, E. J. (2009). Linear viscoelastic models. Part I. Relaxation modulus and melt calibration. Journal of Non-Newtonian Fluid Mechanics 156, 121–128.
2. Borg, T., Pääkkönen, E. J. (2009). Linear viscoelastic models. Part II. Recovery of the molecular weight distribution using viscosity data. Journal of Non-Newtonian Fluid Mechanics 156, 129–138.
3. Bersted, B. H. (1975). An empirical model relating the molecular weight distribution of high-density polyethylene to the shear dependence of the steady shear melt viscosity. *Journal of Applied Polymer Science, 19*, 2167–2177.
4. Bersted, B. H. (1976). A model relating elastic properties of high-density polyethylene melts to the molecular weight distribution. *Journal of Applied Polymer Science, 20*, 2705–2714.
5. Bersted, B. H. (1977). Prediction of dynamic and transient rheological properties of polystyrene and high-density polyethylene melts from the molecular weight distribution. *Journal of Applied Polymer Science, 23*, 1279–1289.
6. Bersted, B. H., & Slee, J. D. (1977). A relationship between steady-state shear melt viscosity and molecular weight distribution in polystyrene. *Journal of Applied Polymer Science, 21*, 2631–2644.
7. Bersted, B. H. (1979). Effect of molecular weight distribution on elongational viscosity of undiluted polymer fluids. *Journal of Applied Polymer Science, 24*, 671–682.
8. Bersted, B. H., Slee, J. D., & Richter, C. A. (1981). Prediction of rheological behavior of branched polyethylene from molecular structure. *Journal of Applied Polymer Science, 26*, 1001–1014.
9. Bersted, B. H. (1985). On the effect of very low levels of long chain branching in rheological behavior in polyethylene. *Journal of Applied Polymer Science, 30*, 3751–3765.
10. Malkin, A., Teishev, A. E. (1991). Flow curve-molecular weight distribution: is the solution of the inverse problem possible. Polymer Engineering Science 31, 1590–1596.
11. Malkin, A. (1990). Some inverse problems in rheology leading to integral equations. Rheologica Acta 29, 512–518.
12. Tuminello, W. H., & Cudré-Mauroux, N. (1991). Determining molecular weight distributions from viscosity versus shear rate flow curves. *Polymer Engineering Science, 31*, 1496–1507.
13. Shaw, M. T., & Tuminello, W. H. (1994). A closer look at the MWD-viscosity transform. *Polymer Engineering & Science, 34*, 159–165.
14. Liu, Y., Shaw, M. T., & Tuminello, W. H. (1998). Obtaining MWD information from the viscosity data of linear polymer melts. *Journal of Rheology, 42*, 453–476.
15. Liu, Y., Shaw, M. T., & Tuminello, W. H. (1998). Optimized data collection for determination of the MWD from the viscosity data of polymer melts. *Polymer Engineering & Science, 38*, 169–176.
16. Liu, Y., & Shaw, M. T. (1998). Investigation of the nonlinear mixing rule for its adequacy in viscosity to molecular weight distribution transforms. *Journal of Rheology, 42*, 267–279.
17. Léonardi, F., Allal, A., & Marin, G. (1998). Determination of the molecular weight distribution of linear polymers by inversion of a blending law on complex viscosities. *Rheologica Acta, 37*, 199–213.
18. Léonardi, F., Allal, A., & Marin, G. (2002). Molecular weight distribution from viscoelastic data: The importance of tube renewal and Rouse modes. *Journal of Rheology, 46*, 209–224.
19. Anderssen, R. S., & Mead, D. W. (1998). Theoretical derivation of molecular weight scaling for rheological parameters. *Journal of Non-Newtonian Fluid Mechanics, 76*, 299–306.

20. Thimm, W., Friedrich, C., Marth, M., & Honerkamp, J. (1999). An analytical relation between relaxation time spectrum and molecular weight distribution. *Journal of Rheology, 43*, 1663–1672.
21. de Vries, S. A., Kale, L. T., Oswald, T., Karjala, T. P. (1996). Molecular structure–Linear viscoelastic properties modeling for constrained geometry catalyst polyolefin resins. In: Aït-Kadi, A., Dealy, J., James, D., Williams, M. (Eds.), Proc. 12th Int. Congr. on Rheology, Quebec, Canada 18–23 August 1996.

Chapter 10
Start-Up and Transient Flow Effects

Start-up and transient shear-stress flows are modelled here using the Unified Model for linear viscoelasticity of polymers. The generated constitutive models rewrite Boltzmann superposition principle and use a *time-rate* separability basis for functions instead of the normally used time–strain separability. This chapter is based entirely on the article published as reference [1].

10.1 Boltzmann Superposition Principle for Transient Viscosity

Transient shear-stresses associated with the shear-stress growth coefficient $\sigma^+(t, \dot{\gamma})$ are complicated to model at different shear rates ($\dot{\gamma}$), as illustrated in Fig. 10.1a. Therefore, we model the time-dependent transient shear viscosity $\eta\,(t, \dot{\gamma})$ at various shear rates. Transient viscosity is a well-known property of many gels and colloids, but is less studied and modelled in polymers, primarily due to the short-period effects and the limitations of instrumentation during measurements. Attempts are often made to avoid this effect during steady-state measurements. Subjecting polymers to high shear rates causes shear thinning as a function of time, while low shear rates can induce shear thickening. The physical background is related to the probability of chain orientation and the degree of entanglement. This transient viscosity can explain some capillary effects and high-speed injection-moulding processes. Start-up stress effects are modelled and verified with measurements in the following section using the generated transient viscosity.

To describe stress and viscosity at different induced shear rates, we use a caret to indicate the effective old viscosity $\hat{\eta}$ at $t = t_0$ developed at a shear rate $(\hat{\eta})$ or $(\hat{\eta}) = \eta(t_0, \dot{\gamma}_0)$, which reaches $\lim_{t\to\infty}\left[\eta(t, \dot{\gamma})\right] = \eta(\dot{\gamma})$ in the steady-state situation. Depending on the shear history, effective viscosity $\hat{\eta}$ at time $t = t_0$ may vary

T. Borg, *Control Theory in Rheology*, https://doi.org/10.1007/978-3-031-88248-7_10

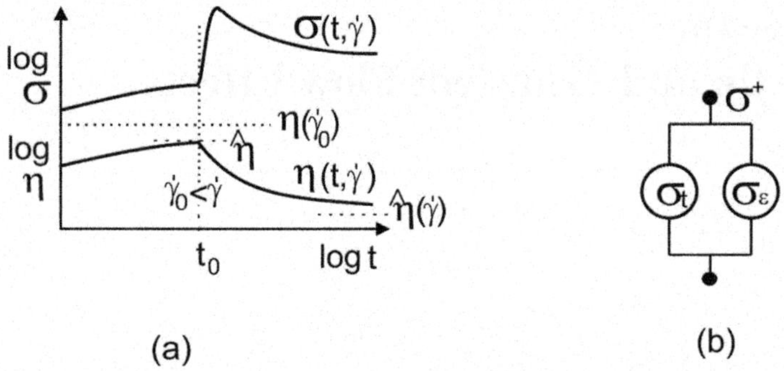

Fig. 10.1 Transient shear-stress and viscosity curves, and schematic of the model for the start-up situation. Borg and Pääkkönen [1]. **a** Graph of time-dependent transient stress $\sigma(t, \dot{\gamma})$ and viscosity $\eta(t, \dot{\gamma})$ during a step increase in the shear rate. The polymer is initially relaxing to steady-state limit $\eta\ (\dot{\gamma}_0)$, when a new shear rate $\dot{\gamma}$ is imposed at t_0 for effective viscosity $\hat{\eta}$. The viscosity curve appears to be significantly simpler than the stress curve. **b** Schematic of the parallel model for stress growth function σ^+ by time-dependent relaxation stress σ_t^+ and secondary elongation stress σ_ε^+ of the shear

considerably from its steady-state limit $\eta(\dot{\gamma}_0)$. In other words, low-quality measurements might yield true effective viscosity $\hat{\eta}$, which can differ from true steady-state viscosity η.

New imposed shear rate $\dot{\gamma}$ results in the viscosity gradually approaching its limit $\lim_{t\to\infty}\left[\eta(t, \dot{\gamma})\right] = \eta(\dot{\gamma})$. The negative or positive viscosity difference $\Delta\eta = \eta - \hat{\eta}$ approaches zero as a function of normalized relaxation $g(t) = G(t)/G_0$. However, during a constant shear all of the molecules do not relax or orientate to every possible direction (or, at least, the effect is minimal in the perpendicular z direction), and thus we need dimension exponent $D \sim 1/3$ since the relaxation of orientation change is in one direction. The time-dependent transient viscosity is added to the effective viscosity $\hat{\eta}$ at time $t = t_0$: $\eta(t, \dot{\gamma}) = \hat{\eta} - \Delta\eta(\mathrm{t})$. At different shear rates the stress results from the continuous loading of the molecules rather than relaxation, and we get

$$\eta(t, \dot{\gamma}) = \hat{\eta} + (\eta(\dot{\gamma}) - \hat{\eta})(1 - g(t))^D \tag{10.1}$$

Now we rewrite the Boltzmann superposition principle for shear viscosity and molecular orientation: the total effect of applying several shear deformations and changes in molecular orientations is simply the sum of their individual effects. Molecular orientation is also related to the effective viscosity, as discussed later. We obtain the general summation from Eq. (10.1) as follows:

$$\eta(t,\dot{\gamma}) = \hat{\eta} + \sum_{i=1}^{N}(1-g(t-t_1))^D \delta\eta(t_1,\dot{\gamma}_1) \tag{10.2}$$

For continuous changes in viscosity and orientation, this sum is generalized to an integral as

$$\eta(t,\dot{\gamma}) = \hat{\eta} + \int_{-\infty}^{t} (1-g(t-\tau))^D \frac{d\eta(\tau)}{d\tau} \tag{10.3}$$

Equations (10.1)–(10.3) can model transient shear viscosity $\eta(t,\dot{\gamma})$ even at high shear rates since $\hat{\eta}$ is used as an absolute starting level with $g(t)$ in a novel manner, where the polymer state can be far from the zero state or starting point.

The Boltzmann superposition principle yields the linear viscoelasticity in a similar way as for all earlier presented formulas for control theory and melt calibration, and we can conclude that we now have linear viscoelasticity starting from the MWD to the transition effects.

The main functions and characteristics of these computations are listed in Table 10.1.

Table 10.1 Main functions and characteristics of computations in this chapter

Fig	In[a]	Out[b]	T[c]	Mw[d]	MwR[d]	Mf[e]	Hf[e]	P' [f]	P'' [f]
10.2	$(\dot{\gamma})$	$\eta(t,\dot{\gamma})$	150	480,000	25.0	20,000	2	0.41	0.18
10.3	$\eta(t,\dot{\gamma})$	$\eta^+(t,\dot{\gamma})\eta$	150	480,000	25.0	20,000	2	0.41	0.18
10.4	$\eta(t,\dot{\gamma})$	$\eta^+(t,\dot{\gamma})\eta$	150	480,000	25.0	20,000	2	0.41	0.18
10.5	$G(t)$	$\eta^-(t)\eta$	150	480,000	25.0	37,800	2.05	0.38	0.20
10.6	$\hat{\eta}_0$	$G(t,\hat{\dot{\gamma}})$	150	480,000	25.0	20,000	2	0.39	0.08
10.7	$\eta(t,\dot{\gamma})$	$\hat{P}\%$	150	240,000	14.6	20,000	2	0.42	0.26
10.8	$\eta(t,\dot{\gamma})$	$\hat{P}\%$	130–250	240,000	14.6	20,000	2	0.32–0.44	0.15–0.27

[a] Viscoelastic function used as a data source
[b] Modelled viscoelastic function
[c] Temperature (°C)
[d] Average molecular weight M_w (g/mol) and polydispersity index M_w/M_n
[e] Structural value *Mf* and conversion factor *Hf*
[f] Elasticity and viscosity values P' and P''

10.2 Shear Stress Growth and Decay Coefficients

In the start-up situation, shear stress growth function $\sigma^+(t,\dot{\gamma}) \equiv \sigma$ and respective shear stress growth coefficient $\eta^+(t,\dot{\gamma}) \equiv {}^{\sigma^+}/_{\dot{\gamma}}$ can be expressed according to nomenclature [4] as approaching its steady-state limit as

$$\lim_{t\to\infty}\left[\eta^+(t,\dot{\gamma})\right] = \eta(\dot{\gamma}) \tag{10.4}$$

The application of a shear stress to a well-relaxed polymer may result in overshoots of stress growth coefficient $\eta^+(t,\dot{\gamma})$ and the molecular elongation forces. There are two types of stresses for stress growth functionσ^+: (1) time-dependent relaxation stress σ_t^+ and (2) the shear rate growth, which generates elongation stress σ_ε^+ for chains before a new orientation and steady state is reached. These forces act in parallel as shown in Fig. 10.1b, and we can write a simple formula for σ^+ that is analogous to parallel resistors in a electric circuit (Fig. 10.3):

$$\frac{1}{\sigma^+} = \frac{1}{\sigma_t^+} + \frac{1}{\sigma_\varepsilon^+} \tag{10.5}$$

In Eq. (10.5) we can use Eq. (10.1) for transient viscosity $\eta(t,\dot{\gamma})$, formulas $\gamma = \dot{\gamma}t$ and $\sigma = \eta\dot{\gamma}$, and simple algebraic conversions such as multiplying by $\dot{\gamma}$ to obtain

$$\frac{\dot{\gamma}}{\sigma^+} = \frac{1}{\eta(t,\dot{\gamma})} + \frac{1}{\eta(t,\dot{\gamma})\dot{\gamma}_\varepsilon t} \tag{10.6}$$

Since elongation stress $\sigma_\varepsilon^+ = \eta_E\dot{\varepsilon}$ is a function of Hencky strain rate $\dot{\varepsilon}$ and Trouton viscosity η_E, we have to convert these into a function of the shear rate. In the linear case $\eta_E\ (\dot{\varepsilon}) = 3\eta(\dot{\gamma})$ and the stress can be written as $\sigma_\varepsilon^+ = \eta_E\dot{\varepsilon} = \eta\dot{\gamma}_\varepsilon$, where the extension shear rate is $\dot{\gamma}_\varepsilon = 3\dot{\gamma}$ and is lower at somewhat higher rates in the non-linear case.

After simple algebraic conversions, this finally yields the formula for start-up flow $\eta^+(t,\dot{\gamma}) \equiv {}^{\sigma^+}/_{\dot{\gamma}}$:

$$\eta^+(t,\dot{\gamma}) = \frac{\eta(t,\dot{\gamma})\dot{\gamma}_\varepsilon t}{\dot{\gamma}_\varepsilon t + 1} \tag{10.7}$$

For shear-stress decay coefficient $\eta^-(t) \equiv {}^{\sigma^-}/_{\dot{\gamma}_\varepsilon}$ and after the cessation of steady shear, we can use classical formula

$$\eta^-(t) = \int_t^\infty G(s)ds \tag{10.8}$$

All of these functions need viscosity measurements as a function of the shear rate to generate the relaxation modulus and wide-ranging viscosity flow. Modelled start-up viscosity curves for IUPAC A LDPE are shown in Fig. 10.4, and the modelled shear-stress decay coefficient is presented in Fig. 10.5.

10.3 Simulations for the Relaxation Modulus

The measurement of relaxation modulus requires at least two steps, including simulating the shear history and the shear of the initial ramp. We rewrote the procedure presented in our earlier papers to find effective viscosity $\hat{\eta}$ and effective zero relaxation modulus $\hat{G}_0$ at $t = t_0$ instead of classical zero modulus G_0. The polymer is no longer fully relaxed after a step strain is used to measure the relaxation modulus, and hence $\hat{G}_0$ is lower than its conventional value G_0. The effective G_0, $\hat{G}_0$, is determined directly by setting $t > 0$ for conventional relaxation modulus $G(t) = \hat{G}_0$ or by using the respective effective shear rate, $\widehat{\gamma}_0$, which gives the respective effective zero viscosity $\hat{\eta}_0 = \eta\left(\widehat{\gamma}_0\right)$. Equation (10.10) can be used to obtain the respective $\hat{G}_0$. The internal effective zero relaxation modulus of the polymer, $\hat{G}_0$, tends to G_0 during a zero shear situation:

$$\lim_{t \to \infty}\left[\hat{G}_0(t)\right] = G_0 \tag{10.9}$$

We used standard simple numerical feedback fitting procedures (e.g. bracketing) while solving Eq. (10.10) by testing the new effective $\hat{G}_0$, values to obtain the formula for effective zero viscosity $\hat{\eta}_0$:

$$\hat{\eta}_0 = \int_{-\infty}^{\infty} t\hat{G}_0 g(t) d \log t \tag{10.10}$$

where $\hat{\eta}_0 = \eta\left(\widehat{\gamma}_0\right)$ is determined by the respective effective shear rate $\widehat{\gamma}_0$. In most cases the polymer still has some orientation and is not in a fully relaxed state, and thus $\eta_0 > \hat{\eta}_0$ with $\widehat{\gamma}_0 > 0$ for the polymer melt, with Eq. (10.11) yielding $\hat{G}_0 < G_0$. All of the shear history can be explained by $\hat{\eta}$ or with the respective effective shear rate $\widehat{\gamma}$ for steady-state viscosity $\eta\left(\widehat{\gamma}\right)$. The effective viscosity is the same as the respective steady-state limit at a constant shear rate or $\dot{\gamma} = \widehat{\gamma}$ and $\hat{\eta} = \eta(\dot{\gamma}) = \eta\left(\widehat{\gamma}\right)$.

Now we use relaxation function $G\left(t, \widehat{\gamma}\right) = G_0(t) - \Delta G_0(t)$, where $G_0(t)$ is the theoretical relaxation modulus for a fully relaxed state, and $\Delta G_0(t) = G_0(t) - \hat{G}_0(t)$ with $\hat{G}_0(t) = \hat{G}_0(t)g(t)$. The principle is similar to that presented for the transient viscosity in Eq. (10.1):

$$G\left(t, \widehat{\gamma}_0\right) = G_0(t) + \left(G_0(t) - \hat{G}_0(t)\right)\left(1 - \frac{G_0(t)}{G_0}\right)^D \tag{10.11}$$

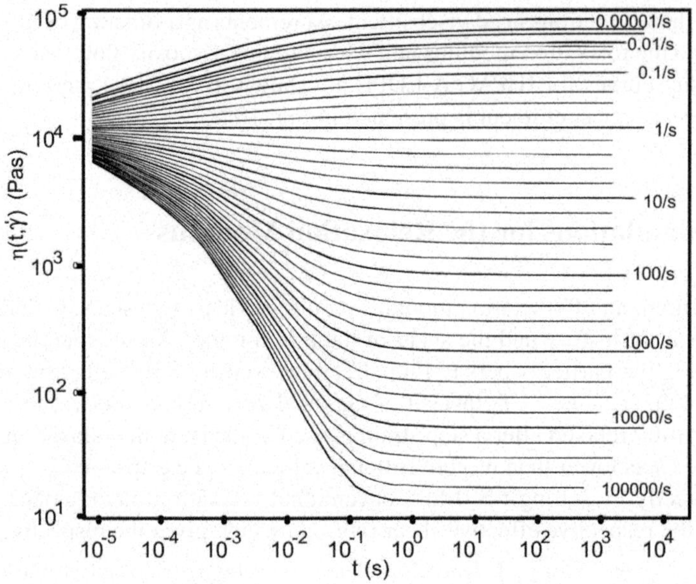

Fig. 10.2 Modelled transient viscosity $\eta(t, \dot{\gamma})$ over a wide range of $\dot{\gamma}$ values for IUPAC A LDPE at 150 °C from the steady and effective shear rate $\dot{\gamma}_0 = \widehat{\dot{\gamma}} = \frac{1}{s}$ at $t = t_0$. The used polymer IUPAC A, a well-characterized low-density polyethylene (LDPE) as described by Meissner [2, 3] and Chap. 3 and elsewhere in this text. Borg and Pääkkönen [1]

Relaxation function $G\left(t, \widehat{\dot{\gamma}}\right)$ can be converted to a standard imposed-ramp relaxation function $G(t, \gamma)$ as a function of strain using the information in Fig. 10.2 if the shear history, implied duration of strain and amount of strain are known.

Most attempts to model $G(t, \gamma)$ using constitutive equations lead to time-strain separability functions [5], which are generally not derived from fundamental principles. The form of these functions provides theoretical simplicity and practical convenience for modelling constitutive equations. Nevertheless, separability is violated in short-time responses immediately after the application of a step strain. Our basis is the new *time-rate* separability shown in Eq. (10.1) for transient viscosity and for the relaxation function in Eq. (10.11). This new approach allows any time window to be modelled even in a highly orientated non-relaxed state.

When a large step strain γ is applied for a long time before relaxation, respective effective shear rate $\widehat{\dot{\gamma}}$ is close to the steady-state shear rate used for deformation. In the opposite case with a high shear rate, the polymer will not have reached the steady state, and relaxation will produce higher forces. Fortunately, Eq. (10.1) yields effective shear rate $\widehat{\dot{\gamma}}$ and viscosity $\hat{\eta}$ to $G\left(t, \widehat{\dot{\gamma}}\right)$ at any time, and converts it further to $G(t, \gamma)$.

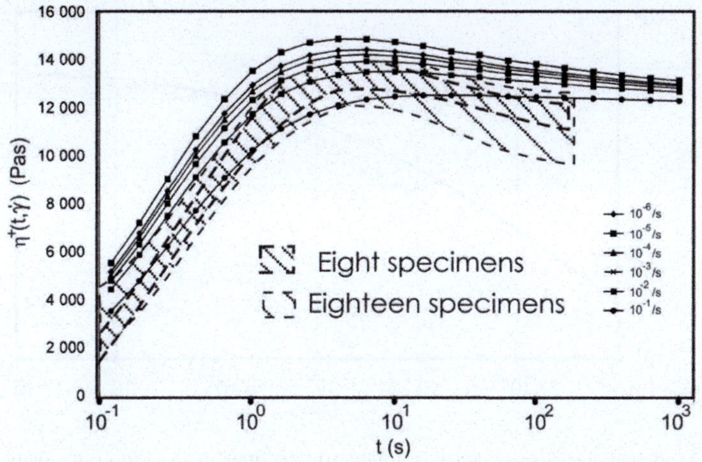

Fig. 10.3 Measured and modelled start-up $\eta^+(t, \dot{\gamma})$ viscosity curves of IUPAC A LDPE at 150 °C. The specimen was measured at 18 laboratories (wide hatching), and the results from the most reliable specimens (narrow hatching) are close to the modelled data up to 100 s. The discrepancy for t > 100 s is attributable to measurement limitations and distortions induced by the applied strains. Borg and Pääkkönen [1]

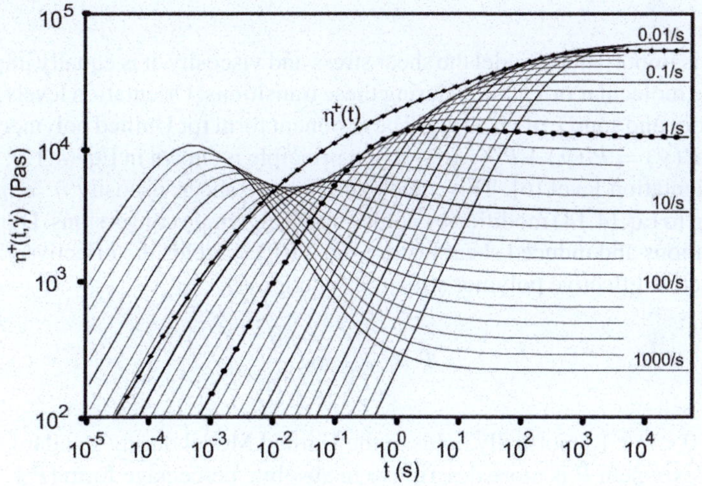

Fig. 10.4 Modelled start-up $\eta^+(t, \dot{\gamma})$ viscosity curves of IUPAC A LDPE at new induced lower and higher shear rates. The lower thick line corresponds to the induced $\dot{\gamma} = 1$/s shear rate, as shown in Fig. 10.3. The upper thicker line was calculated from the relaxation modulus using the classical equation. Borg and Pääkkönen [1] (Fig. 10.5)

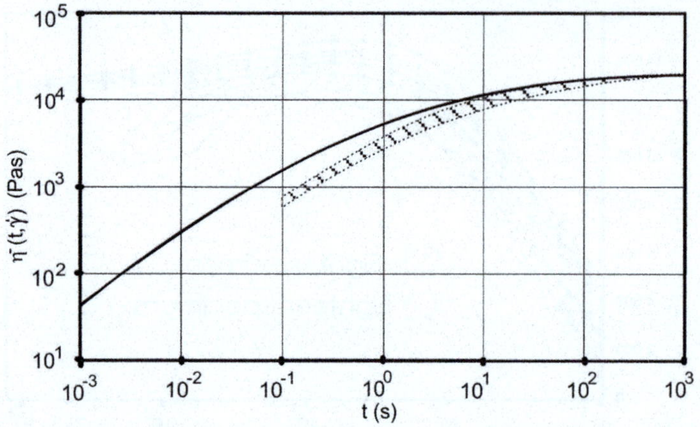

Fig. 10.5 Modelled shear-stress decay coefficient $\eta^{-}(t)$ after the cessation of steady shear. The measurements of IUPAC A LDPE at 150°C are indicated by the hatched area. Borg and Pääkkönen [1]

10.4 Simulating the Molecular Orientation During Transition Effects

Whilst it is important to model the shear stress and viscosity, it is equally important to model the molecular orientation during these transitions. Orientation levels represent the memory and source of viscosity. The exponent form for Unified polymer structure function $P(\dot{\gamma}) = P'(\dot{\gamma}) + P''(\dot{\gamma})$, which in principle is shown in Eq. (4.12), is related to the orientation level [6]. This yields an effective shear viscosity $\hat{\eta} = \eta_0 |\dot{\gamma}|^{-P(\dot{\gamma})}$ according to Eq. (4.14) modified by some simple algebraic conversions. During shear rate variations and induced shear history, Eq. (10.12) yields the effective viscosity $\hat{\eta}$ for respective effective polymer structure function $\hat{P}$:

$$\hat{P} = -\frac{1}{\dot{\gamma}} \log \frac{\hat{\eta}}{\eta_0} \tag{10.12}$$

Since $0 < \hat{P} < 1.5$ normally holds in the Unified Model during standard flows, the effective exponent $\hat{P}$ is more descriptive in absolute percentage form ($\hat{P}\% = 100\,\hat{P}$) or even in relative form: (Fig. 10.6)

$$\hat{P}_{REL}\% = \frac{100\hat{P}}{\hat{P}_{MAX}} \text{ scale} \tag{10.13}$$

Since the modelled and measured MWDs are related to molecular orientations at different deformations, we can use the principle to model orientation during flows and for final end products, where a high degree of orientation can cause warping and dimensional deviations over time. This would simultaneously provide accurate and

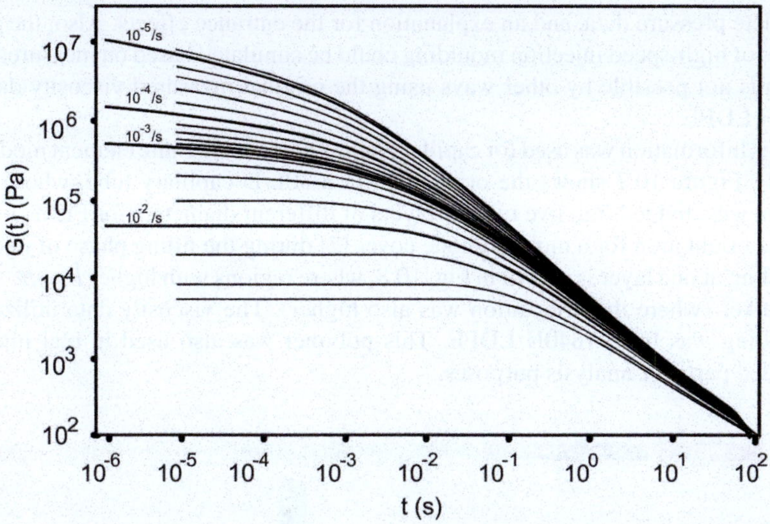

Fig. 10.6 Simulated relaxation modulus $G\left(t, \hat{\gamma}\right)$ at different effective shear rates related to effective zero relaxation modulus $\hat{G}_0$ and also for more familiar $G(t, \gamma)$ tests. The used polymer is IUPAC A. Borg and Pääkkönen [1]

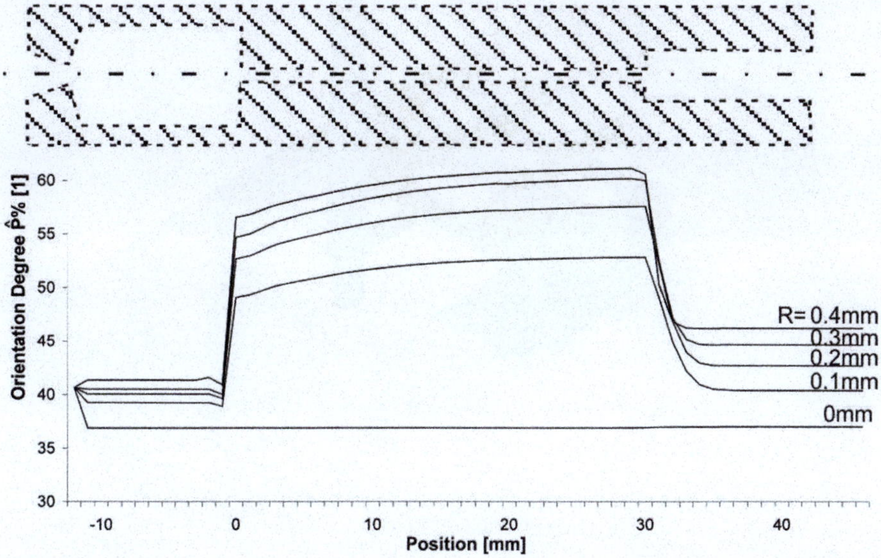

Fig. 10.7 Orientation simulation for different layers in a 30L/D capillary for 1840H LDPE. Borg and Pääkkönen [1]

complete pressure data, and an explanation for the entrance effects. Also, the filling phases of high-speed injection moulding could be simulated based on measured data, which is not possible by other ways using the original measured viscosity data for 1840H LDPE.

This information was used for capillary simulations using finite element modelling (FEM). Figure 10.7 shows the orientation in a 30L/D capillary tube, whose cross section was divided into five tubular layers of different diameters for FEM.

The orientation for a mobile-phone cover [7] during the filling phase of injection moulding at one layer is shown in Fig. 10.8, where regions with higher $\hat{P}_{REL}\%$ values are darker (where the orientation was also higher). The viscosity data utilized for modelling was from 1840H LDPE. This polymer was also used in real injection moulded parts for analysis purposes.

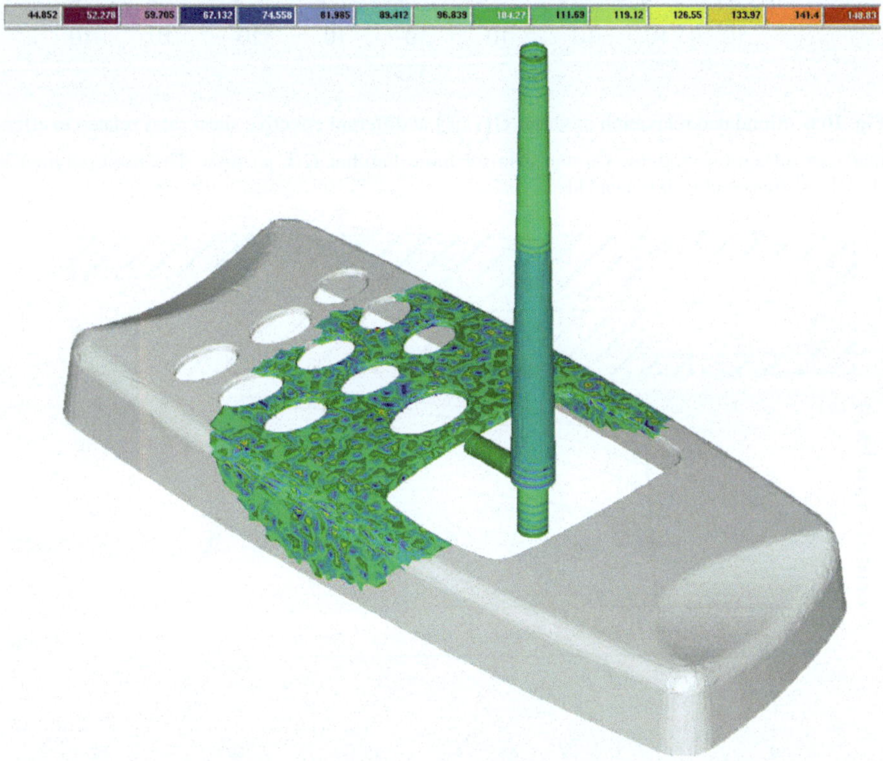

Fig. 10.8 Orientation simulation layer at $t = 0.1$ s during the filling of an injection mould for a mobile-phone cover. Different coloured regions and respective $P =$ values/100 indicate orientation levels in the filled melt for 1840H LDPE. Eq. (10.12) for P is expressed in common logarithmic form, while the formula used in the figure is in natural logarithmic form; therefore, the real values shown in the figure are 0.434*P. It appeared that the melt flow was labile at high shear rates on the order of at least 10,000/s. Borg and Pääkkönen [1]

The filling was complete in 0.3 s, and there was nonhomogeneous turbulence in the orientation at $t = 0.1$ s during the filling phase (as shown in Fig. 10.8). It appeared that the melt flow was labile at high shear rates on the order of at least 10,000/s. The orientation results were similar at different filling rates, but relaxation was evident close to the front surface. Further differences in the orientation level were evident in different layers, less in the top and bottom layers and more in the middle layers. The presence of a high residual orientation level in a completed cover can result in warping, and such pieces cannot be used in the final phone assembly due to dimensional deviations.

References

1. Borg, T., Pääkkönen, E. J. (2009). Linear viscoelastic models. Part III. Start-up and transient flow effects from the molecular weight distribution. *Journal of Non-Newtonian Fluid Mechanics, 159*, 17–25.
2. Meissner, J. (1975). Basic parameters, melt rheology, processing and end-use properties of three similar low-density polyethylene samples. *Pure and Applied Chemistry, 42*, 551–662.
3. Meissner, J. (1972). Modifications of the Weissenberg Rheogoniometer for measurement of transient rheological properties of molten polyethylene under shear. Comparison with tensile data. *Journal of Applied Polymer Science, 16*, 2877–2899.
4. Dealy, J. M. (1995). Official nomenclature for material functions describing the response of a viscoelastic fluid to various shearing and extensional deformations. *Journal of Rheology, 39*, 253–265.
5. Kwon, Y., & Cho, K. S. (2001). Time-strain nonseparability in viscoelastic constitutive equations. *Journal of Rheology, 45*, 1441–1452.
6. Ianniruberto, G., & Marrucci, G. (2001). A simple constitutive equation for entangled polymers with chain stretch. *Journal of Rheology, 45*, 1305–1318.
7. Borg, T., Keinonen, M., Poranen, P., Pääkkönen, E. J. (2004). In: Lee, J. W., Lee, S. J. (Eds.), Time dependency of viscosity in the flow simulations, *Proc. XIVth Int. Congr. on Rheology*, The Korean Society of Rheology, Seoul, Korea, 22–27, CR24.

Chapter 11
Elongational Viscosity

We first present a formula for uniaxial elongational viscosity without strain hardening. This formula is derived in a manner similar to the methods discussed in previous chapters, starting from control theory principles and leading to the Unified Model.

The next step in our exploration involves incorporating the effective strain-hardening distribution, which, in many cases, is related to phenomena like long-chain branching. This chapter is based on the published article [1].

11.1 Introduction

It is beyond the scope of this article to provide a complete summary of all the group studies of elongational viscosity; more information on this is available in the historical reviews by Petrie [2] (for the last 100 years) and Mackley [3] (for the last 40 years). In recent decades, measurements have been made by Meissner and Hostettler [4], Laun and Schuch [5], Münstedt [6], Münstedt et al. [7, 8], Sentmanat [9], Hassager et al. [10], Aho et al. [11] and van Ruymbeke et al. [12, 13]. Fewer studies have modelled the elongational viscosity, among them being Wagner [14], Wagner et al. [15], Rolón-Garrido and Wagner [16], Rolón-Garrido et al. [17] with a generalization of the theory of Doi and Edwards [18, 19], Laun [20] on memory function, Rauschenberger and Laun [21] on recursive model, McLeish and Larson [5] and Inkson et al. [22] on "pom-pom" constitutive equations [23], Likhtman and Graham [24] on the Roli-poly model, Auhl et al. [25] on the volume of tube and van Ruymbeke et al. [12] by Mixing law and tube pressure. A good new composition of field is by Münstedt [26] in *Elastic Behavior of Polymer Melts*.

Our development process was conducted in the opposite manner to earlier studies. This unconventional approach ensures an accurate, simple, yet consistent model with a strong foundation in physics for these complex phenomena and flows. In other

T. Borg, *Control Theory in Rheology*, https://doi.org/10.1007/978-3-031-88248-7_11

words, we can progressively delve into more detailed and intricate microstructures while maintaining the model's consistency. The strength of this approach lies in the fact that the complexity of the polymer chain structure is not a limitation for developing robust semi-empirical viscosity models. The greater challenge arises from studying the microstructure of the chains.

Several new equations for different flows, without relying on artificial relaxation time schemas, are presented in the earlier chapters. Since this principle starts from a clean slate and introduces a new perspective in rheology, the developments cannot be assessed using Occam's razor. In fact, only a few material-dependent constants are used. It was recently discovered that the rheologically effective distribution (RED), which correlates very accurately and linearly with viscoelastic properties, is actually a melt chromatogram, similar to the eluent, elution curve, or elugram in liquid chromatography methods.

11.2 Elongational Viscosity Without Strain Hardening Effect

We can directly write the formula in the start-up for uniaxial elongational viscosity $\overset{\sim+}{\eta}_{E}(t)$ using control theory and Unified Model in a similar way to previous work on the relaxation modulus and the dynamic and shear viscosities. The strain hardening effects coming from long-chain branches (LCBs) or similar structures are not yet included in to the elongational viscosity $\overset{\sim+}{\eta}_{E}(t)$. As previously for the case in the start-up of shear viscosity being a function of shear and elongation, we have to use our new time–rate separability basis (which is new to rheology) when deriving functions for the parallel-resistor analogy discussed in the Sect. 7.2. Now the elongation is again a function of one rate variable, $\dot{\varepsilon}$, and hence we do not need to start from steady-state elongational viscosity $\eta(\dot{\varepsilon}) \equiv \frac{\sigma}{\dot{\varepsilon}}$—which is a function of constant Hencky strain rate $\dot{\varepsilon} = \frac{\mathrm{d}\ln(L/L_0)}{\mathrm{d}t}$ and constant tensile stress σ. Moreover, the steady-state elongational viscosity or "Trouton viscosity" is difficult to measure.

We use an alternative strategy to get elongational viscosity by imposing net tensile stress σ_E monitoring as a function of time. At first we model $\eta_E^+(t)$ without any strain-hardening effects or influences of LCBs.

Thus, we write a formula directly for elongational viscosity $\overset{\sim+}{\eta}_{E}(t)$, where the tilde mark over the "η" indicates that the strain-hardening effect is not yet included:

$$\log \frac{\tilde{\eta}_E^+(t)}{\tilde{\eta}_0^+} = -\log \frac{t}{t_c} \int_{-\infty}^{\log t} \left(Pw(\log \tau) + P'' \right) \mathrm{d} \log \tau \tag{11.1}$$

where $\overset{\sim+}{\eta}_{0}$ is the elongational viscosity at characteristic time $t_c = 1/\mathrm{s}$ similar way as earlier. We developed the formula similar way starting from control theory and

Unified Models as for shear viscosity $\eta(\dot{\gamma})$ by Eq. (4.7), complex viscosity $\eta^*(\omega)$ by Eq. (5.8) and relaxation modulus $G(t)$ by Eq. (6.11). The rheologically effective distribution (RED) $w(\log t)$ relates not only to orientated but also stretched chains that lose their entanglements. RED function $w(t)$ can be regarded as respective chromatogram or elugram distribution with liquid chromatography of GPC/SEC. This RED, presenting here as an elastic component, influences the polymer normally after $t > 0.001$ s during elongation. Since the possible measurement range starts after $t > 0.01$ s, we do not need to use the $w''(\log t)$ component of the RED″ distribution including viscoelastic effects, since it is always at normalized value $w'' = 1$ in these measurement ranges and hence can be simplified to $P'' w''(\log t) = P''$. In most cases $P > 0$, but $P'' < 0$.

A simple relations can be written for the melt calibration, $M(t)$, as a function of time, where the value of Mf is M at $t_1 = 1/\text{s}$ as follows:

$$w(\log M) = w(\log t)\Big|_{t=t_1\left(\frac{Mf}{M}\right)^{Hf}} \tag{11.2}$$

$$\tilde{w}(\log t) = w(\log M)\Big|_{M=Mf\left(\frac{t}{t_1}\right)^{-\frac{1}{Hf}}} \tag{11.3}$$

Before doing nothing for $\tilde{w}(\log t)$ we have to normalize to get useable rheologically effective distribution RED $w(\log t)$ for computations.

Melt calibration has close similarities with the widely used linear broad-standard calibration with SEC, which approach requires a broad standard of known number-average M_n and weight-average $\overline{M}w$ molecular weights. The use of different Mf and Hf values allows the REDs to be converted into their respectively MWDs, where Mf sets the absolute molecular weight and Hf sets the polydispersity value.

The Analytical formula for viscosity $\overset{\sim+}{\eta}_E(t)$ can be written by applying a procedure similar to that used in our previous studies, but this was not required for the present study.

11.3 Strain Hardening with a Monotonically Increasing Effect

Elongational viscosity $\overset{\sim+}{\eta}_E(t)$ includes basic strain-hardening component originally $w_H(\log \varepsilon)$ as a function of time t and elongation rate $\dot{\varepsilon}$ or effective strain-hardening distribution $w_H(\log t, \dot{\varepsilon})$, where $t = \varepsilon/\dot{\varepsilon}$ with monotonically increasing behaviour, at least in the absence of the overshoot effect, as has been done in most other simulation models.

Long-chain branches (LCBs) and high-molecular-weight (HMW) end fractions influence strain-hardening effects during the initial stages of elongation experiments. The origin of this additional response lies in the oriented and tautly stretched chains between entanglements that have not yet had time to disentangle and relax. The longest chains are the first to exhibit this tautly stretched characteristic, leading to strong strain hardening as they have no more free loops available. Eventually, the flow becomes non-homogeneous on a micro or even semi-micro scale, which accelerates the breakdown of elongation experiments.

Since hardening is related to the length of chains, LCBs and HMW end fractions, we set the respective hardening and rheologically effective distribution $\mathrm{RED_H}$ or $w_{\mathrm{H}}(\log t, \dot{\varepsilon})$ (note that we are using the "H" subscript, even though the origin can also be other than LCBs or HMW end fractions). The strain-hardening effect is adjusted by the value of P_{H}. Now we can write the complete formula for start-up of elongational viscosity $\eta_{\mathrm{E}}^{+}(t)$ as

$$\log\frac{\eta_{\mathrm{E}}^{+}(t,\dot{\varepsilon})}{\eta_0^{+}} = -\log\frac{t}{t_{\mathrm{c}}}\int_{-\infty}^{\log t}\left(Pw(\log\tau) + P'' + P_{\mathrm{H}}(\dot{\varepsilon})\,w_{\mathrm{H}}(\log\tau,\dot{\varepsilon})\right)\mathrm{d}\log\tau \quad (11.4)$$

For developing RED $w_{\mathrm{H}}(\log t, \dot{\varepsilon})_{\mathrm{H}}$ we use its own $\mathrm{RED_H}$ according to Eq. (11.1) with its own *Mf* and *Hf* factors, since this rheological behaviour is totally different.

From the standpoint of molecular dynamics and the tube model described in Chap. 12, we obtain that statistical unit segment length L_{s} increases and diameter D_{s} decrease relatively for a unit backbone during elongation. Elasticity function $P\prime = k\prime\frac{D_{\mathrm{s}}}{L_{\mathrm{s}}}$ and especially P_{H} decrease in a similar way in Eq. (11.4) giving less strain hardening effect.

For the used elongation rates, with $\dot{\varepsilon}$ <1/s, meaningful differences to $\eta_{\mathrm{E}}^{+}(t)$ values are not observed at the beginning (i.e., $t < 0.1$ s), but the amount of strain hardening depends on the used rate described by strain-hardening coefficient $\chi(t, \dot{\varepsilon})$ [27, 28]. They discuss the molecular structure, referring to long-chain branching. But no simulation model is presented for absolute measured values of $\chi(t, \dot{\varepsilon})$, thus we have to develop a simple model to perform modelling and obtain simulation results. The strain-hardening effect, $P_{\mathrm{H}}(\dot{\varepsilon})$, is linearly related to strain-hardening coefficient $\chi(t, \dot{\varepsilon})$. Normalized distributions $w(\log t)$ and $w(t, \dot{\varepsilon})$ and strain hardening effects during different strain rates are shown in Fig. 11.1.

The above-described procedure for simulations is rather insensitive to the actual form of the used MWD, which make it possible to use MWD data copied from the literature or to use nominal distributions. The results are much more sensitive to the actual form of the long-chain branching distribution (LCBD) and $w_{\mathrm{H}}(\log\tau, \dot{\varepsilon})$.

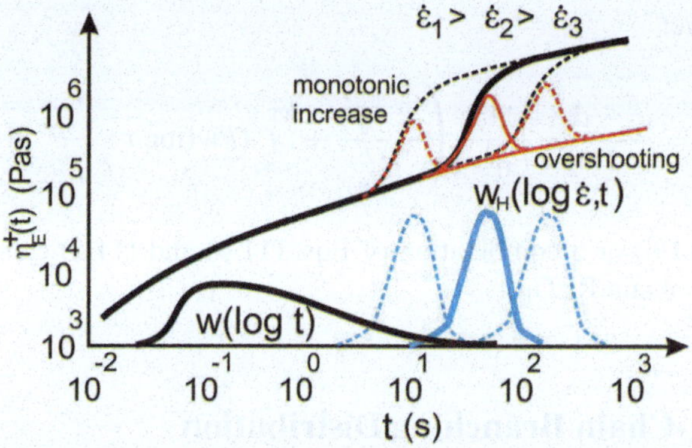

Fig. 11.1 Elongational viscosity $\eta_E^+(t)$ with LCBs giving strain hardening, and a schematic model for converting REDs and RED$_H$s into the respective MWDs. RED$_H$s are scaled with their effectiveness by $P_H(\dot{\varepsilon})$ values or $P_H(\dot{\varepsilon})w_H(\log t, \dot{\varepsilon})$, where $\varepsilon = \dot{\varepsilon}t$ at different elongation rates according to Eq. (11.4) for a monotonic increase and Eq. (11.5) for the overshoot strain-hardening effects. Borg and Pääkkönen [1]

11.4 Strain-Hardening Component with the Overshoot Property

The elongational viscosity at large strains has been studied less, as has strain-hardening viscosity. Rasmussen et al. [29] measured the viscosity at large Hencky strains of up to 6–7 for low-density polyethylene (LDPE) as shown in Fig. 11.4 and linear low-density polyethylene (LLDPE). They observed that LLDPE behaves as shown in Fig. 11.6 and according to Eq. (11.5), whereas LDPE exhibits a viscosity overshoot effect that appears as a bump. We thus rewrite Eq. (11.4) for the overshoot-type strain hardening as follows:

$$\log\frac{\eta_E^+(t)}{\eta_0^+} = -\log\frac{t}{t_c}\left(\int_{-\infty}^{\log t}(Pw(\log\tau) + P'')\mathrm{d}\log\tau + P_H(\dot{\varepsilon})w_H(\log\tau, \dot{\varepsilon})\right) \quad (11.5)$$

The form of the left side of strain-hardening distribution $w_H(\log t, \dot{\varepsilon})$ is similar to that at the beginning of strain hardening from Eq. (11.4) or (11.5). However, the final form of the subsequent strain hardening remains unclear.

11.5 Effective Strain-Hardening Distribution RED$_H$

We extract strain-hardening distribution RED$_H$ from Eq. (11.4) using a procedure similar to that described for the shear viscosity in the Chap. 4, complex viscosity in the Chap. 5 or relaxation modulus in the Chap. 6. We obtain $w_H(\log t, \dot{\varepsilon})$ by deriving as follows from the measured differences to $\eta_E^+(t)$ to accurately obtain the shape of the $\mathrm{RED_H}$ curve:

$$w_{\mathrm{H}}(\log t, \dot{\varepsilon}) = -\frac{\mathrm{d}}{\mathrm{d}\log t}\frac{1}{P_{\mathrm{H}}(\dot{\varepsilon})}\left(\frac{\ln\frac{\eta_{\mathrm{E}}^{+}(t)}{\eta_0^{+}}}{\ln\frac{t}{t_{\mathrm{c}}}} + \int_{-\infty}^{\log t}\left(Pw(\log\tau) + P''\right)\mathrm{d}\log\tau\right) \quad (11.6)$$

Equation (11.6) or a combination of Eqs. (11.4) and (11.5) can be solved in a similar way to obtain $\mathrm{RED_H}$.

11.6 Long-Chain Branching Distribution

The strain-hardening distribution $\mathrm{RED_H}$ is converted to a direct relation of the long-chain branching distribution (LCBD) through melt calibration using a simple relational formula. Real distributions for LCBDs are very rarely presented alongside MWDs in the literature.

A nuclear magnetic resonance method (NMR) is used to detect LCB, but is known that NMR can be used to detect side chain branches with up to 8 carbon atoms only by Fleury et al. [30]. Many alternative ways have been developed to obtain a single numerical value (Malmberg et al. [31]; Costeux [32]; Chen et al. [33]; Wood-Adams et al. [34, 35]; Sugimoto et al. [36]; Vega et al. [37]) including by utilizing the Zimm–Stockmayer equation (Zimm and Stockmayer 1949). Several rheological methods are suitable for identifying LCB (Vega et al. [38]; Gabriel and Münstedt [39]; Malmberg et al. [40, 41]; Wood-Adams and Dealy [34]; Gabriel and Münstedt [39]), such as van Gurp–Palmen (van Gurp and Palmen [42]; Trinkle et al. [43]) and Cole–Cole (García-Franco et al. [44]) plots, flow activation energy (Keßner et al. [45]), thermorheology (Keßner et al. [46]), the dynamic modulus (Sugimoto et al. [36]), and relaxation times, but all suffer from the problem of not being able to convert the obtained results into absolute numerical values.

We employ a different procedure using a more realistic one-dimensional LCBD with the total number of inner backbones. In Fig. 11.2a on the left is shown used terminology [47] for the primary chain as backbone, directly linked chains are free arms and next connected chains are inner backbone segments. Of course this LCBD in Fig. 11.2b can be fractionalized later on to bivariate MW–LCB distribution if needed (Keßner et al. [45]). Main distribution $w_{\mathrm{MAIN}}(M)$ can be detected rheologically according to Chap. 9 or by GPC/SEC, which is used mainly for this study and known also insensitive to detect LCBs. For showing final results we use sum of RED or MWD distributions to get normalized sum value one as follows: $\int w_{\mathrm{MAIN}} + \int w_{\mathrm{LCB}} = 1$.

LCB distribution as a function of M or $w_{\mathrm{H}}(M) = w_{\mathrm{H}}(\log t, \dot{\varepsilon})$, where $\dot{\varepsilon} = 1/\mathrm{s}$ used here, is gained by the right selection for Eq. (11.2) by their own values Mf and Hf for melt calibration similar way as are done with calibration curve and GPC/SEC procedures. As constants P and P_{H} set effects for viscoelastic properties and they are inside log function, we get simple formula for strain-hardening coefficient $\chi(\dot{\varepsilon})$

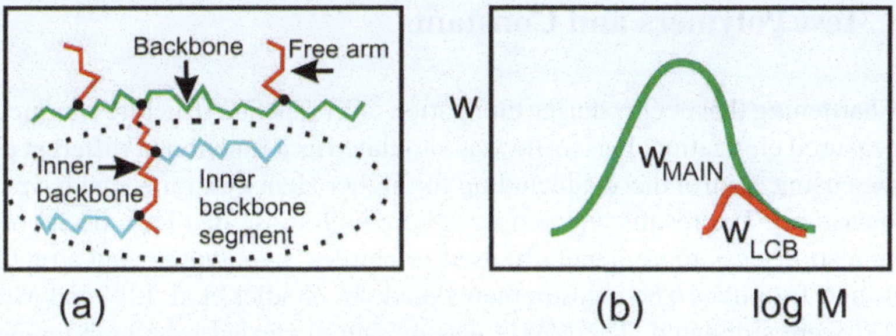

Fig. 11.2 Schematic diagram of detecting LCBD $w_{\mathrm{LCB}}(M)$. **a** Molecule chain has backbone, free arms and inner backbone segments. **b** From GPC/SEC or by using narrow dynamic viscoelastic data range to get normalized MWD or $w_{\mathrm{MAIN}}(M)$. From strain-hardening distribution $w_{\mathrm{H}}(\log t, \dot{\varepsilon})$ and own *Mf* and *Hf* values for melt calibration we get absolute LCBD or $w_{\mathrm{LCB}}(M)$ distribution. Borg and Pääkkönen [1]

$= \log(P_{\mathrm{H}}(\dot{\varepsilon})/P)$ as both constants in Eq. (11.5) are inside logarithm. The influence of right size for final LCB distribution $w_{\mathrm{LCB}}(M)$ by using relation $1/\chi$ of normalized MWD and LCBD as follows

$$w_{\mathrm{LCB}}(M) = \frac{1}{M_{\mathrm{b}}} \log\left(\frac{P_{\mathrm{H}}}{P}\right) w_{\mathrm{H}}(M) \tag{11.7}$$

where M_{b} is molar mass m of a backbone element, for PE $M_{\mathrm{b}} = 14$ mol/g. In principle LCBD can be outside of MWD even though measured by GPC/SEC, as these methods do not detect LCBs. Thus we get absolute LCBD or $w_{\mathrm{LCB}}(M)$ by using different structural value *Mf* and conversion factor *Hf* for conversion from $\mathrm{RED_H}$.

We can renew the formula for the LCB frequency—expressed as the number of long chain branches per 1,000 carbon atoms as presented in previous studies [32, 38]—using the above results for LCBD as follows:

$$\mathrm{LCB}/1000\ \mathrm{C} = 1000 M_{\mathrm{b}} \sum \frac{w_{\mathrm{LCB}}(M_{\mathrm{i}})}{M_{\mathrm{i}}} \tag{11.8}$$

where w_{LCB} is the weight fraction of LCBD, M_{i} is the MW of each fraction, and M_{b} is the molar mass of a backbone element [$M_{\mathrm{b}} = 14$ for polyethylene (PE)]. Complete weight fraction w_{i} of the MWD and the average number of branches per molecule were originally used in the formula (Chen et al. [33]). However, now this value consists not only of branches on the backbone but also of inner backbones, as shown in Fig. 11.2a. A major disadvantage of these types of formulas is that the average MW of backbones (and now branches) strongly influence the obtained values.

11.7 Test Polymers and Constants

Strain-hardening that occurs during elongation of a branched structure was modelled. The measured elongational viscosity was simulated as a function of different elongation rates using control theory, including for higher Hencky strains that have not yet been measured. The results are compared here with those that have been observed, and we also discuss more generally used principles, hypotheses, and structures of models and formulas. The measurements made by Stadler et al. [28] and Münstedt et al. [8] were simulated. The MWD was measured carried by a high temperature GPC coupled with a multiangle light scattering apparatus (MALLS). MWD for LDPE come from RheoPower Database. As we know, GPC is not sensitive to detect LCBs. Elongational viscosity simulations were performed by executing a Unified Model of control theory on a standard PC with the commercial RheoPower software package using the RheoDeveloper program with experimental elongation moduli.

We did not need to use the more sensitive RheoAnalyzer program for MWDs, since their effects on the rheological properties are less than those of LCBDs in this application. Our description of results starts by using functional formulas of Eq. (11.1) giving accurate real fits with measurements by adjusting REDH, then is shown and discussed different strategies for modelling elongational viscosity at high Hencky strains. Along calculation are some words of LCB average chain length and distribution and branching intensity. One bivariate RED–REDH distribution chart as a function of time and rate is shown and finally MW-LCB distribution results converted to two apparent MWD charts.

11.8 Procedure

The elongational viscosity was modelled first without and then with strain hardening. Since the presented formulas are new, and hence the values of constants are unknown, we explain the step-by-step best-fit routine and by Fig. 11.3 based on the measurements and our model starting from control theory. We did not use Eq. 11.6 to detect $w_{\mathrm{H}}(\log t, \dot{\varepsilon})$ in this study, since we fitted our curves with previously published measured data. We obtained that manually copied data to derive by Eq. 11.6 was sensitive to errors. The elongational viscosity is first developed without the strain-hardening effect, which involves the following steps:

1. Draw a known MWD in the RheoDeveloper program and save it in a database. Set *Mf* and *Hf* constants as a-priori information using values found previously for the complex viscosity.
2. Set first test P', P'' and zero $\tilde{\eta}_0^+$ values for the polymer sample, and compute and compare them with measured elongation values. Adjust the values so that the simulation result on the output chart is as close as possible to images imported from measured data.

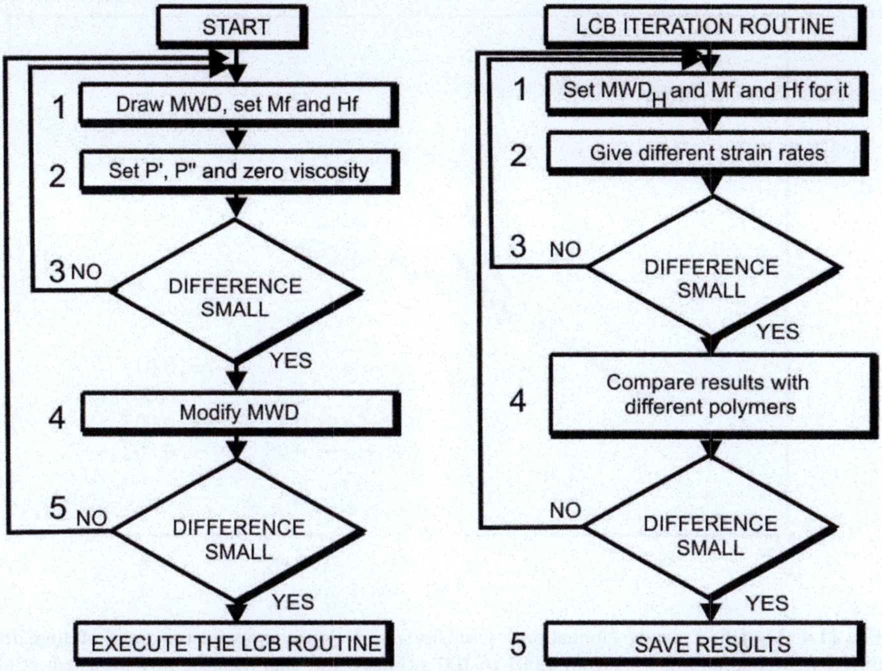

Fig. 11.3 Flowchart of the simulation procedure. The data can be processed further on a spreadsheet. Borg and Pääkkönen [1]

3. Alter the *Mf* and *Hf* values and repeat step 2.
4. Modify and alter a little-known MWD to see if errors between the modeled and measured results increase so as to check if the used MWD is acceptable.
5. Test different types of polymer samples in order to ascertain the correct values for constants. Repeat the above steps as many times as necessary to obtain satisfactory results, at which point you can proceed to the next level.

The following LCBD iteration routine involves the strain-hardening function and distribution:

1. Set separate *Mf* and *MwR* values for hardening distribution RED_H $w_\text{H}(\log t, \dot{\varepsilon})$ with strain hardening constant $P_H > 0$.
2. Obtain some elongation rates $\dot{\varepsilon}$ and compare them with measurements. If necessary, select new *Mf* and *Hf* constants for $w_\text{H}(\log t, \dot{\varepsilon})$.
3. Check the values with different polymers and attempt the obtain the best fit by using different Mf, Hf, $\eta_0^+(t)$, $P_H(\dot{\varepsilon})$ and $\dot{\varepsilon}$ values.
4. You can try to get better fit by drawing manually hardening distribution RED $w_\text{H}(\log t, \dot{\varepsilon})$.
5. Save the final results in ASCII format in log-0.2 steps on a wide timescale and on the strain rate matrix.

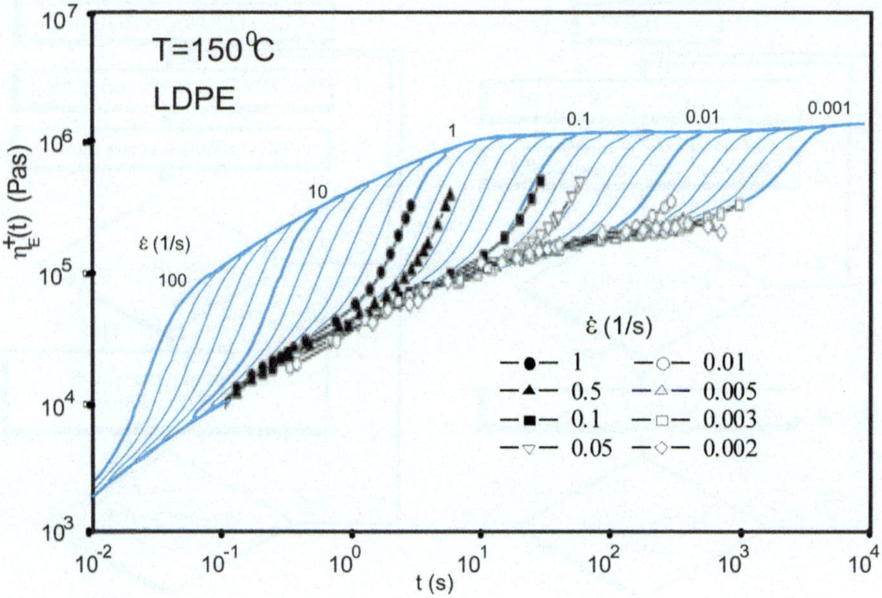

Fig. 11.4 Modelled steady elongational viscosity $\eta_E^+(t)$ flow curves as functions of time for different elongation rates ($\dot{\varepsilon}$) from 0.001 to 100 s (lines), and data measured by Münstedt et al. [8] for LDPE (symbols). Modelled rates on decades (e.g., 10/s, 1/s and 0.1/s) are indicated by thicker lines. Borg and Pääkkönen [1]

The data can be processed further on a spreadsheet such as LibreOffice Calc or Excel so as to combine MWD and LCBD. Measured data can be joined in CAD program with the modelled and simulated results.

11.9 Used Main Characteristics

A polymer structure is modelled by the values test P', P'' and zero viscosity η_0^+, polymer structural value Mf and conversion factor Hf between scales. The values of P' and P'' obtained from the viscosity fitting procedure are listed in Table 11.1, and no ad hoc constants or values were used. A constant temperature of T = 150 °C was used for LDPE and LLDPE, while T = 180 °C was used for polypropylene (PP).

For the effective strain-hardening distribution LCBD we used constant value Hf = 0.6 to allow comparisons between different samples. The used Mf values might not be accurate, but it is within the correct range and of course its value differs at least for PP. The main characteristics of the samples are listed in Table 11.1.

Table 11.1 Main characteristics of the samples

	LDPE	LLDPE	PP 2	CSTR-LDPE 1	CSTR-LDPE 2
Mw[a]	213,000	102,000	574,000	4,167,000	2,271,000
MwR[a]	12.9	3.2	9.3	21.4	14.3
Mf[b]	70,000	30,000	200,000	70,000	70,000
Hf[b]	2.05	2.05	4	2.05	2.05
η_0^{+}[c]	4.6	4.68	4.1	4.87	4.18
P'[d]	0.42	0.65	0.82	0.22	0.51
P"[d]	–0.70	–0.70	–0.80	–0.68	–0.68
P_H[e]	2.6	2	2.6	4	3.4
Mw LCBD[f]	10,640,000	9,670,000	7,150,000	17,030,000	9,670,000
MwR LCBD[f]	5.3	2.7	1.4	6.5	2.7
Mf[g]	4,000,000	4,000,000	4,000,000	40,000,000	40,000,000
X	1.8	1.1	1.5	2.9	1.9
LCB/1000 *C*[h]	0.022	0.012	0.034	0.001	0.002

[a] The used *Mw* (g/mol) and *MwR* values may differ slightly from those measured using GPC/SEC
[b] Polymer structural value *Mf* (g/mol) and conversion factor between scale *Hf*
[c] Elongational viscosity (Pas) at $t = 1$/s
[d] Obtained elasticity P' and P'' viscosity values
[e] Strain-hardening constant P_H at rate $\dot{\varepsilon} = 1$/s
[f] *Mw* (g/mol) and *MwR* values obtained for the effective strain-hardening distribution
[g] Polymer structural value *Mf* (g/mol) for strain-hardening distribution, $Hf = 0.6$
[h] Strain-hardening coefficient $\chi(\dot{\varepsilon})$ at rate $\dot{\varepsilon} = 1$/s
[i] Long chain branching frequency per 1000 C carbon atoms

11.10 Uniaxial Elongational Viscosity

We model uniaxial elongational viscosity as a function $P_H(\dot{\varepsilon})$ according to measurements, but at a much higher Hencky strain (ε) compared to that obtained in practical measurements. We first simulate a modern LDPE (Lupolen 1840 H) for well known and classical IUPAC A, as done in previous Chapters. The elongation viscosity results are shown in Fig. 11.4 and used RED_H functionals with $P_H(\dot{\varepsilon})$ in Fig. 11.5 according to Eq. (11.5). We observe that simulation results accurately fit the measurements made by Münstedt et al. [8]. Since the elongation software moduli of the RheoDeveloper program was still an experimental version, the output of the elongation viscosity was still in log-0.2 steps, which corresponds to the following elongation rates: 0.01/s, 0.016/s, 0.025/s, 0.04/s, 0.06/s, 0.1/s, 0.16/s, 0.25/s, 0.40/s, 0.6/s and 1/s. On the other hand, elongation rates of 0.01/s, 0.1/s and 1/s were used during measurements, whereas the data were for 0.002/s, 0.003/s, 0.03/s, 0.05/s and 0.5/s, which fall between the modelled elongation rates.

As steady-state elongation viscosity is sometimes thought to run through a maximum as a function of elongation rate, we can simulate this by using smaller strain-hardening constants P_H at higher elongation rates. Used products for Fig. 11.4

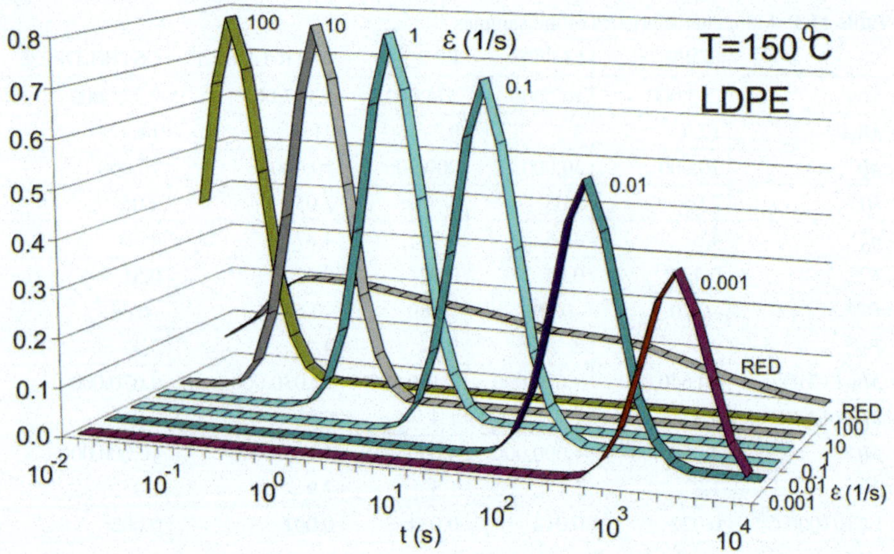

Fig. 11.5 Used $P_H(\dot{\varepsilon})\, w_H(t, \dot{\varepsilon})$ products for generating Fig. 11.4 by product of $P_H(\dot{\varepsilon})$ and effective strain-hardening distribution RED_H or $w_H(t, \dot{\varepsilon})$ shown only on decade steps of $\dot{\varepsilon}$ for clarify. On the back is shown segment of used RED on time scale related to MWD. Borg and Pääkkönen [1]

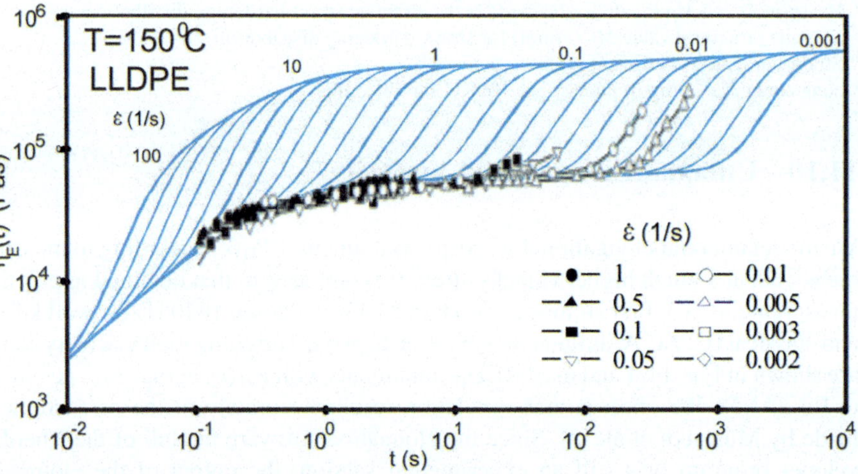

Fig. 11.6 Measured and modelled elongational viscosities for LLDPE. The stresses are orders of magnitude less than for LDPE in Fig. 11.4. Borg and Pääkkönen [1]

this time variable $P_H(\dot{\varepsilon})$ as a function of rate and effective strain-hardening distribution RED_H or $\mathrm{P}_H(\dot{\varepsilon})\ \mathrm{RED}_H$ can be seen on Fig. 11.5.

We have three important structures: the average LCB chain length and its distribution. Thirdly, the amount of LCB branches has no doubt a lot of effect. We can obtain that REDHs related to branches has much shorter and sharper effect on the elongation viscosity than RED related to the backbones of chains. The phenomenon indicates that longer chain lengths of LCBD have more influence as shorter ones behaving similar way as in melt calibration principle. It is clear that the modelled curves accurately fit the measurements, with some differences being due to the use of different elongation rates during measurements and simulations, and at low rates ($\dot{\varepsilon} < 0.01/s$) due to relaxation of taut-tied chains.

11.11 Modelled Elongational Viscosity at High Hencky Strains

We now try to understand the mechanism underlying the observed and measured strain-hardening effects and the accuracy of the extrapolations for high Hencky strains. Strain-hardening, which occurs during film blowing or sheet foaming, is a useful and wanted feature in many practical plastic manufacturing processes. This property influences film casting and coating processes and it also gives the final end product a kind of self-healing mechanism. On the other hand it has drawbacks with LCBs in that it represents a source of strong and long-duration shrinkage that can cause the end products to warp. Strain-hardening has also been exploited as a useful feature for heat-shrinkable sleeves, as described in the US patent by Borg [48].

As discussed above, strain-hardening is sensitive to LCBs and HMW end fractions, which we attribute to long taut-tied stretched chains between entanglements not having sufficient time to disentangle and relax, causing inhomogeneity on a micro or even a semimicro scale due to smaller chain bundles combining. Burghelea et al. [49] measured a significant difference between the normally used integral form for viscosity and the locally measured elongational viscosity. They discussed geometric non-uniformity and its relation to the stress maximum. Although we are studying macromolecules here, the longest LCB can still have an oriented length of many microns, which can induce local inhomogeneity in smaller bundles of molecules that causes breakdown during elongation measurements. Fibrilar structure and morphology development of blends during and after elongational deformation were observed by Starý and Münstedt [50]. The elongation experiment with the Hencky strain variable with a strongly diminishing cross-section differs from many other procedures, and the local inhomogeneity in macromolecular chains can rapid break down the elongation. On the other hand, large cross-sections produce inaccurate measurements. Elongation experiments are difficult to carry out at the laboratory scale, but they are very important for understanding the behaviour and structure of polymers. Steady macro-scale plastic manufacturing processes may achieve much higher Hencky strains, and thus it might be necessary to model the complete rate range and even strains outside achievable measurement ranges. Measured and modelled elongational viscosities for LLDPE are shown in Fig. 11.6. The stresses are orders of magnitude less than for LDPE in Fig. 11.4. This time we have used constant value for strain-hardening value P_H as the data of measurements at higher rates.

Comparison of Figs. 11.4 and 11.6 reveals that strain hardening has a greater effect at higher elongation rates for LDPE and at lower elongation rates for LLDPE. Several possible mechanisms could underlie the strain-hardening of LLDPE, including the presence of a few LCBs or components with high molar masses, which are not detectable by classical analytical methods, and phase separation in the molten state (Lohse et al. [51]). Strain-hardening may peak at various times of elongation measurements, which has been studied recently by Stadler et al. [28]. Since the underlying mechanism remains somewhat obscure, we did not try to model this feature but instead used the best fit with maximal strain-hardening and the same level for all elongation rates.

The next puzzle is the possibility of overshoot during elongation. Rasmussen et al. [29] obtained an overshoot for LDPE Lupolen 1840 D, whereas Burghelea et al. [50] argued that this phenomenon does not exist. This discrepancy is why we present all of our simulation results as monotonically increasing without an overshoot effect in charts for visual clarity; moreover, although the amount of overshoot is not known, we still believe that some overshoot is present (as discussed above).

In summary, the present measurements accurately provided the left side of the effective hardening distribution REDH or LCBD, where this procedure is sensitive to the results. The level of strain-hardening was obtained by fitting procedure.

11.12 Simulations of Polymers with a High Level of Hardening

Polymers with a high level of hardening and LCB were simulated. For comparison we show a different chemical type of polymer in Fig. 11.7. The simulated data were the measurements of the highly branched PP made by Münstedt et al. [8] and used for all results for demonstration purposes constant strain-hardening P_{H} value, but uncertain results at higher elongations are dashed in the figure.

The last two presented simulations are of high molecular weight LDPEs known to have a very high LCB content: (1) including many LCBs polymerized at the laboratory scale in a continuously stirred tank reactor (CSTR) and (2) the polymer designated as CSTR-LDPE 1. The results are shown in Fig. 11.8. We did not know whether or not overshoot was present or the form of LCB distribution, which is why we used Eq. (11.5) without the overshoot effect in the monotonically increasing form for visual clarity (as discussed above). One novel interesting feature of this simulation was that its results were consistent with the measured strain-hardening spectrum for rates (i.e., $\dot{\varepsilon}$ values) of 0.1/s and 0.01/s, but not for rates of $\dot{\varepsilon} = 10$/s and

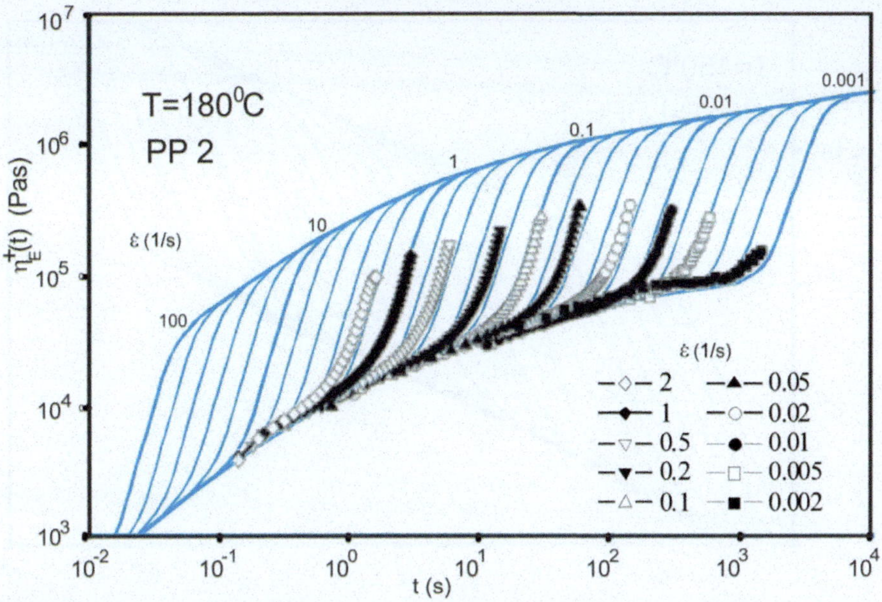

Fig. 11.7 Simulated results using Eq. 11.5 for highly branched PP according to measurements by Münstedt et al. [8] (sample designated as PP 2 in that report). Strain-hardening occurs over a wide range of elongation rates. Since the measurements do not show any overshoot, a large-strain approximation is used according to Eq. 11.5. We used constant strain-hardening P_H to simulate approximations in this case. Borg and Pääkkönen [1]

0.001/s. The results for another polymer, designated CSTR-LDPE 2 and which has fewer LCBs, are shown in Fig. 11.9. The simulation results for this polymer were consistent with the measured strain-hardening spectrum for all rates. Moreover, the hardening effect was not as strong as in Fig. 11.8.

High-molecular-weight LDPEs known to have a very high LCB content: (1) including many LCBs polymerized at the laboratory scale in a continuously stirred tank reactor (CSTR) and (2) the polymer designated as CSTR-LDPE 1. The results are shown in Fig. 11.6. We did not know whether or not overshoot was present or the form of LCB distribution, which is why we used Eq. (11.5) without the overshoot effect in the monotonically increasing form for visual clarity (as discussed above). Strain hardening has also been exploited as a useful feature for heat-shrinkable sleeves, as described in the US patent by Borg [49].

The results for another polymer, designated CSTR-LDPE 2 and which has fewer LCBs, are shown in Fig. 11.7. The simulation results for this polymer were consistent with the measured strain-hardening spectrum for all rates. Moreover, the hardening effect was not as strong as in Fig. 11.8.

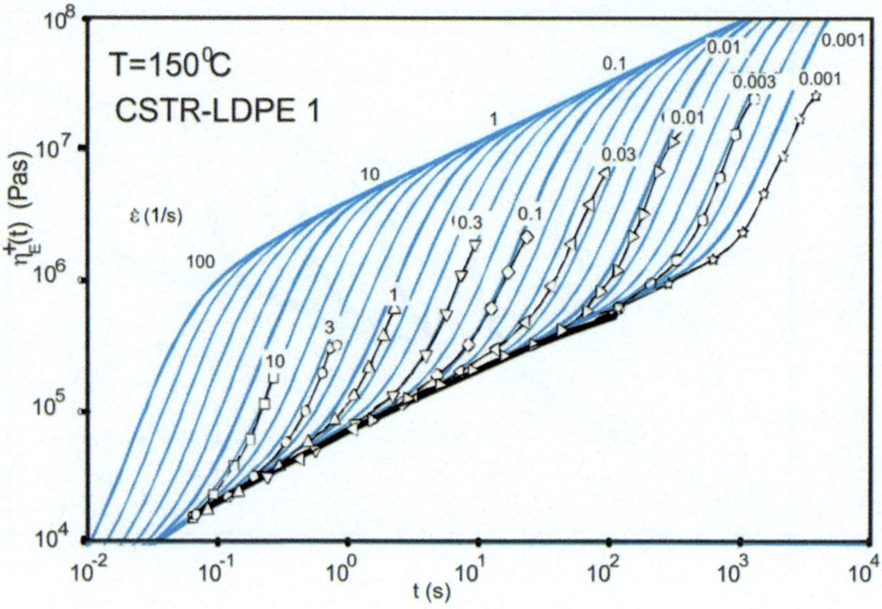

Fig. 11.8 Simulated and measured spectra for CSTR-LDPE 1, with a high content of LCBs. Borg and Pääkkönen [1]

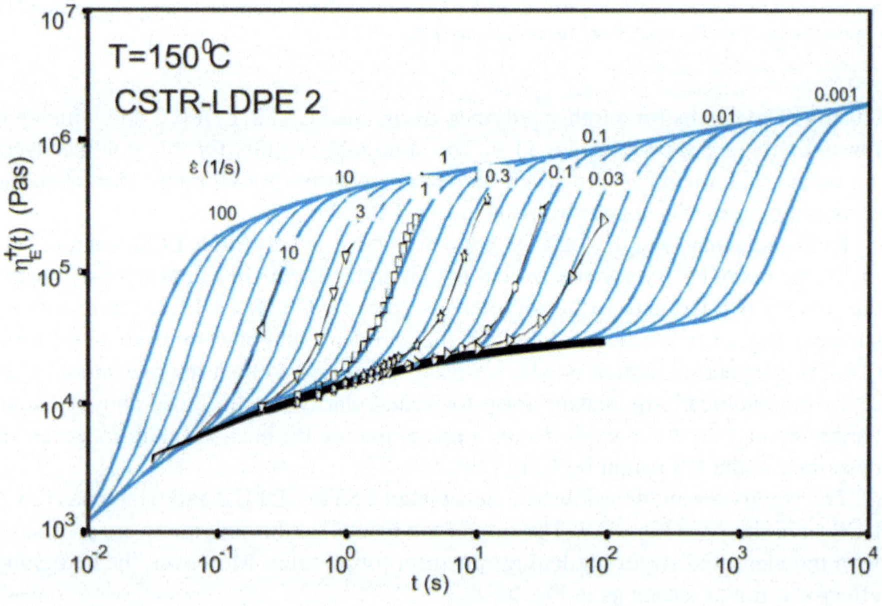

Fig. 11.9 Simulated and measured spectra for CSTR-LDPE 2, with a lower content of LCBs. Borg and Pääkkönen [1]

11.13 Simulated MW-LCB Distributions

We generate long-chain branching distribution (LCBD) with MWD by melt calibration (Eq. 11.2) and by relation formula (Eq. 11.7). RED_Hs were obtained by a best-fit routine on a step-by-step basis based on the measurements and presented formulas. During the procedure it was found that the results were very sensitive even to the local forms of RED_Hs, which meant that we had to draw the final form of the distribution curve manually in many cases. Since the shape of the LCB distribution is the same on a logarithmic scale as the original RED_Hs, the results provide valuable information on polymer structure and the constituent LCBDs.

We selected for conversions by melt calibration for $w_H(\log t, \dot{\varepsilon})$, where $\dot{\varepsilon} = 1$/s. We depicted in Fig. 11.5 accurate RED–RED_Hs chart, now we present strain-hardening distributions converted by melt calibration (Eq. 11.2) to LCBD $w_H(\log M)$ on M scale.

The $w_H(\log t, \dot{\varepsilon})$ strain-hardening distributions are scaled by their strain-hardening constant P_H and respective main normalized MWDs in order to help readers to evaluate the final bimodal MWD distributions in Figs. 11.10 and 11.11. Figure 11.10 presents the MWDs and LCBDs used for commercial LDPE, LLDPE and PP.

Figure 11.11 depicts MWDs and LCBDs for LDPE formed at the laboratory scale and containing a large amount of LCBs. The figure shows similarities with data obtained by GPC/MALLS and the LCBs detected using our new method.

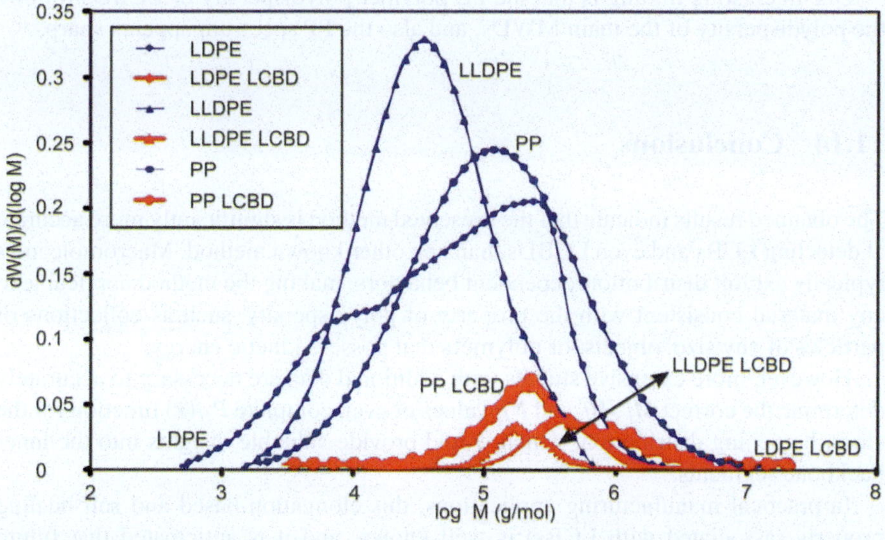

Fig. 11.10 The used MWDs and LCBDs, obtained from effective strain-hardening distribution RED_H or $w_H(\log t, \dot{\varepsilon})$ by Eq. 11.8 for LDPE, LLDPE and PP simulations. Normalized MWD for LLDPE and PP were extracted from GPC/SEC measurements and LDPE from RheoPower database. Borg and Pääkkönen [1]

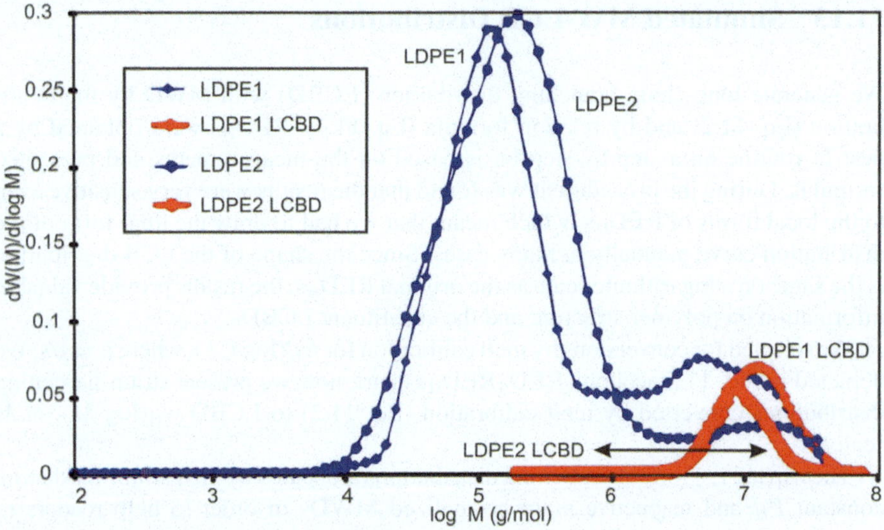

Fig. 11.11 The used MWDs and LCBDs, obtained from effective strain-hardening distribution RED_H or $w_\text{H}(\log t, \dot{\varepsilon})$ by Eq. (11.8) for LDPE simulations. MWDs are from GPC/MALLS measurements. Best-fit procedures revealed similarities between LCBDs as observed previously. Borg and Pääkkönen [1]

One interesting finding is that the PE polymer polydispersity of LCBDs follows the polydispersity of the main MWDs, and also the PP spectrum appears sharp.

11.14 Conclusions

The obtained results indicate that the presented method is significantly more accurate at detecting LCBs and even LCBDs than any other known method. Macromolecules typically exhibit distribution-dependent behaviors, making the method applicable to any material consistent with the property of polydispersity, such as collections of particles of any size, objects, or polymers that possess kinetic energy.

However, more extensive studies with additional data are necessary to accurately determine the correct Mf, Hf and P_H values or even complete $P_H(\dot{\varepsilon})$ function for the strain-hardening distribution, which would provide valuable insights into the inner backbone segments.

In practical manufacturing applications, this elongation-based and self-healing property (associated with LCBs) is well known, and it is anticipated that future advancements in measuring instruments will allow this phenomenon to be fully described and analysed.

References

1. Borg, T., & Pääkkönen, E. J. (2011). Linear viscoelastic model for elongational viscosity by control theory. *Rheologica Acta, 51*, 371–384.
2. Petrie, C. J. S. (2006). One hundred years of extensional flow. *Journal of Non-Newtonian Fluid Mechanics, 137*, 1–14.
3. Mackley, M. (2010). Stretching polymer chains. *Rheologica Acta, 49*, 443–458.
4. Meissner, J., & Hostettler, J. (1994). A new elongational rheometer for polymer melts and other highly viscoelastic liquids. *Rheologica Acta, 33*, 1–21.
5. Laun, H. M., & Schuch, H. (1989). Transient elongational viscosities and drawability of polymer melts. *Journal of Rheology, 33*, 119–175.
6. Münstedt, H. (1979). New universal extensional rheometer for polymer melts. *Journal of Rheology, 23*, 421–436.
7. Münstedt, H., Kurzbeck, S., & Egersdörfer, L. (1998). Influence of molecular structure on rheological properties of polyethylenes part II. *Elongational Behavior, Rheol Acta, 37*, 21–29.
8. Münstedt, H., Kurzbeck, S., & Stange, J. (2006). Importance of elongational properties of polymer melts for film blowing and thermoforming. *Polymer Engineering & Science, 46*, 1190–1195.
9. Sentmanat, M. (2004). Miniature universal testing platform: From extensional melt rheology to solid-state deformation behaviour. *Rheologica Acta, 43*, 657–669.
10. Hassager, O., Marin, J. M. R., Yu, K., & Rasmussen, H. K. (2010). Polymeric liquids in extension: Fluid mechanics or rheometry? *Rheologica Acta, 49*, 543–554.
11. Aho, J., Rolón-Garrido, V. H., Syrjälä, S., & Wagner, M. H. (2010). Measurement technique and data analysis of extensional viscosity for polymer melts by Sentmanat extensional rheometer (SER). *Rheologica Acta, 49*, 359–370.
12. van Ruymbeke, E., Nielsen, J., & Hassager, O. (2010). Linear and nonlinear viscoelastic properties of bidisperse linear polymers: Mixing law and tube pressure effect. *Journal of Rheology, 54*, 1155–1172.
13. van Ruymbeke, E., Muliawan, E. B., Hatzikiriakos, S. G., Watanabe, T., Hirao, A., & Vlassopoulos, D. (2010). Viscoelasticity and extensional rheology of model Cayley-tree polymers of different generations. *Journal of Rheology, 54*, 643–662.
14. Wagner, M. H. (1979). Zur Netzwerktheorie von Polymer-Schmelzen. *Rheologica Acta, 18*, 33–50.
15. Wagner, M. H., Kheirandish, S., Stange, J., & Münstedt, H. (2006). Modeling elongational viscosity of blends of linear and long-chain branched polypropylenes. *Rheologica Acta, 46*, 211–221.
16. Rolón-Garrido, V. H., & Wagner, M. H. (2007). The MSF model: Relation of nonlinear parameters to molecular structure of long chain branched polymer melts. *Rheologica Acta, 46*, 583–593.
17. Rolón-Garrido, V. H., Pivokonsky, R., Filip, P., Zatloukal, M., & Wagner, M. H. (2009). Modelling elongational and shear rheology of two LDPE melts. *Rheologica Acta, 48*, 691–697.
18. Doi, M., & Edwards, S. F. J. (1978). Dynamics of concentrated polymer systems. Part 2. Molecular motion under flow. *Journal of the Chemical Society, Faraday Transactions, 74*, 1802–1817.
19. Doi, M., & Edwards, S. F. J. (1979). Dynamics of concentrated polymer systems. Part 4. Rheological properties. *Journal of the Chemical Society, Faraday Transactions, 75*, 38–54.
20. Laun, H. M. (1978). Description of the non-linear shear behaviour of a low density polyethylene melt by means of an experimentally determined strain dependent memory function. *Rheologica Acta, 17*, 1–15.
21. Rauschenberger, V., & Laun, H. M. (1997). A recursive model for Rheotens tests. *Journal of Rheology, 41*, 719–737.
22. McLeish, T. C. B., & Larson, R. G. (1998). Molecular constitutive equations for a class of branched polymers: The pom-pom polymers. *Journal of Rheology, 42*, 81–110.

23. Inkson, N. J., McLeish, T. C. B., Harlen, O. G., & Groves, D. J. (1999). Molecular constitutive equations for a class of branched polymers: The pom-pom polymers. *Journal of Rheology, 43*, 873–896.
24. Likhtman, A. E., & Graham, R. S. (2003). Simple constitutive equation for linear polymer melts derived from molecular theory: Rolipoly equation. *Journal of Non-Newtonian Fluid Mechanics, 114*, 1–12.
25. Auhl, D., Chambon, P., McLeish, T. C. B., & Read, D. J. (2009). Elongational flow of blends of long and short polymers: Effective stretch relaxation time. *Physical Review Letters, 103*, 136001.
26. Münstedt, H. (2019). *Elastic behavior of polymer melts*. Carl Hanser Verlag.
27. Kurzbeck, S., Oster, F., & Münstedt, H. (1999). Rheological properties of two polypropylenes with different molecular structure. *Journal of Rheology, 43*, 359–374.
28. Stadler, F. J., Kaschta, J., Münstedt, H., Becker, F., & Buback, M. (2009). Influence of molar mass distribution and long-chain branching on strain hardening of low density polyethylene. *Rheologica Acta, 48*, 479–490.
29. Rasmussen, H. K., Nielsen, J. K., Bach, A., & Hassager, O. (2005). Viscosity overshoot in the start-up of uniaxial elongation of low density polyethylene melts. *Journal of Rheology, 49*, 369–381.
30. Fleury, G., Schlatter, G., & Muller, R. (2004). Non linear rheology for long chain branching characterization, comparison of two methodologies: Fourier transform rheology and relaxation. *Rheologica Acta, 44*, 174–187.
31. Malmberg, A., Gabriel, C., Steffl, T., Münstedt, H., & Löfgren, B. (2002). Long-chain branching in metallocene-catalyzed polyethylenes investigated by low oscillatory shear and uniaxial extensional rheometry. *Macromolecules, 35*, 1038–1048.
32. Costeux, S. (2003). Modeling of randomly branched polymers produced by combination of several single-site catalysts: Toward optimization of melt properties. *Macromolecules, 36*, 4168–4187.
33. Chen, X., Costeux, C., & Larson, R. G. (2010). Characterization and prediction of long-chain branching in commercial polyethylenes by a combination of rheology and modeling methods. *Journal of Rheology, 54*, 1185–1206.
34. Wood-Adams, P. M., & Dealy, J. M. (2000). Using rheological data to determine the branching level in metallocene polyethylenes. *Macromolecules, 33*, 7481–7488.
35. Wood-Adams, P. M., Dealy, J. M., DeGroot, A. W., & Redwine, O. D. (2000). Effect of molecular structure on the linear viscoelastic behavior of polyethylene. *Macromolecules, 33*, 7489–7499.
36. Sugimoto, M., Suzuki, Y., Hyun, K., Ahn, K. H., Ushioda, T., Nishioka, A., Taniguchi, T., & Koyama, K. (2006). Melt rheology of longchain-branched polypropylenes. *Rheologica Acta, 46*, 33–44.
37. Vega, J. F., Expósito, M. T., Martínez-Salazar, J., Lobón-Poo, M., Barcina, J. O., Martínez, A. G., & López, M. (2011). Molecular architecture and linear viscoelasticity of homogeneous ethylene/styrene copolymers. *Rheologica Acta, 50*, 207–220.
38. Vega, J. F., Fernándes, M., Santamaría, A., Muñoz-Escalona, A., & Lafuente, P. (1999). Rheological criteria to characterize metallocene catalyzed polyethylenes. *Macromolecular Chemistry and Physics, 200*, 2257–2268.
39. Gabriel, C., & Münstedt, H. (2002). Influence of long-chain branches in polyethylenes on linear viscoelastic flow properties in shear. *Rheologica Acta, 41*, 232–244.
40. Malmberg, A., Liimatta, J., Lehtinen, A., & Löfgren, B. (1999). Characteristics of long chain branching in ethene polymerization with single site catalysts. *Macromolecules, 32*, 6687–6696.
41. Malmberg, A., Kokko, E., Lehmus, P., Löfgren, B., & Seppälä, J. (1998). Long-chain branched polyethene polymerized by metallocene catalysts t[Ind]2ZrCl2/MAO and Et[IndH4]2ZrCl2/MAO. *Macromolecules, 31*, 8448–8454.
42. van Gurp, M., & Palmen, J. (1998). Time temperature superposition of polymeric blends. *Rheology Bulletin, 67*, 5–8.

43. Trinkle, S., Walter, P., & Friedrich, C. (2002). Van Gurp–Palmen Plot II—classification of long chain branched polymers by their topology. *Rheologica Acta, 41*, 103–113.
44. García-Franco, C. A., Srinivas, S., Lohse, D. J., & Brant, P. (2001). Similarities between gelation and long chain branching viscoelastic behavior. *Macromolecules, 34*, 3115–3117.
45. Keßner, U., Kaschta, J., & Münstedt, H. (2009). Determination of method-invariant activation energies of long-chain branched low-density polyethylenes. *Journal of Rheology, 53*, 1001–1016.
46. Keßner, U., Kaschta, J., Stadler, F. J., Le Duff, C. S., Drooghaag, X., & Münstedt, H. (2010). Thermorheological behavior of various short- and long-chain branched polyethylenes and their correlations with the molecular structure. *Macromolecules, 43*, 7341–7350.
47. Costeux, S., Wood-Adams, P., & Beigzadeh, D. (2002). Molecular structure of metallocene-catalyzed polyethylene: Rheologically relevant representation of branching architecture in single catalyst and blended systems. *Macromolecules, 35*, 2514–2528.
48. Borg, T. (Muolon Oy), Method of producing a shrinkable product U.S. Patent 5,053,174, October 1 1991.
49. Burghelea, T. I., Starý, Z., & Münstedt, H. (2011). On the viscosity maximum during the uniaxial extension of a low density polyethylene. *Journal of Non-Newtonian Fluid Mechanics, 166*, 1198–1209.
50. Starý, Z., & Münstedt, H. (2008). Morphology development in PS/LLDPE blend during and after elongational deformation. *Journal of Polymer Science Part B: Polymer Physics, 46*, 16–27.
51. Lohse, D. J., Milner, S. T., Fetters, L. J., Xenidou, M., Hadjichristidis, N., Mendelson, R. A., García-Franco, C. A., & Lyon, M. K. (2002). Welldefined, model long chain branched polyethylene. 2. Melt rheological behaviour. *Macromolecules, 35*, 3066–3075.

Chapter 12
Temperature and Viscoelastic Relations

Viscoelasticity and temperature dependencies are explained through molecular dynamics and Control Theory, which lead to the development of the Control Theory (CT) and Unified Models. Molecular dynamics related to various flow behaviours and Unified formulas have been modelled and discussed in previous sections. Temperature transformations modelled for flows are discussed in this chapter, summarizing from the original paper [1]. An Analytical model for temperature dependence as a rheologically effective distribution (RED) for temperature (RED^T) has been developed. This model is directly related to the thermal expansion curve and the tube diameter as functions of temperature, which are crucial in our model for dynamic chains based on the dumb-bell model [2] and friction.

Our previous studies revealed the great puzzle that the molecular weight distribution (MWD) or the respective RED computed from the relaxation modulus or complex and shear viscosity is not temperature sensitive, although these curves are clearly strongly temperature-dependent. Figure 12.1 shows that the curves of relaxation modulus $G(t)$ and complex viscosity $\eta^*(\omega)$ at 250 and 130 °C are totally different, but the related elastic REDs (RED's) have minimal differences. Different curves were generated using elastic P' and viscous P'' constants and deviation in zero viscosity η_0 and modulus G_0. The figure shows only the foremost distributions, and frequency log ω approximately varies according to 0.13%/1 °C on a logarithmic scale.

The above puzzle conflicts with many previous studies on rheology and time–temperature superposition (TTS), whereas similar types of results have been obtained by van Gurp and Palmen [3] and many other authors [4–10]. However, since our results obtained using precise models were more reliable than those developed previously, we were able to investigate this anomaly and verify TTS only for thermorheologically simple materials.

We developed compact and Analytical formulas using a principle similar to that used in our previous studies. Some measurements at different temperatures, new computations, and the respective WLF and Arrhenius equations are compared, which

T. Borg, *Control Theory in Rheology*, https://doi.org/10.1007/978-3-031-88248-7_12

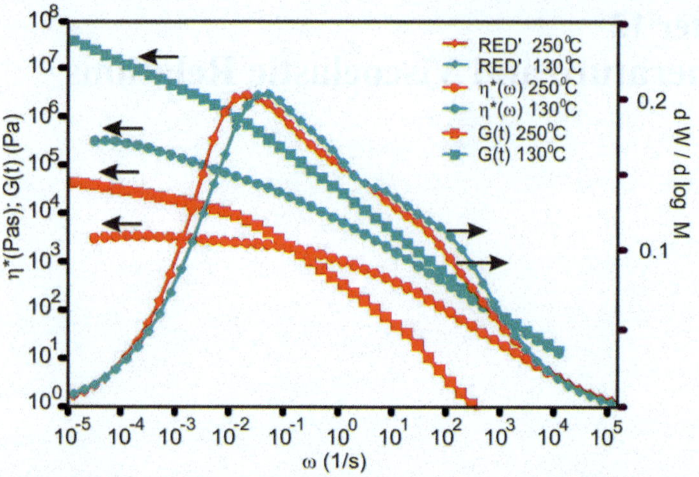

Fig. 12.1 Chart visualizes the puzzle that the elastic RED' and its related MWD computed by melt calibration are not temperature sensitive, whereas the shown foremost RED's at 250 and 130 C are linearly related to $G(t)$ and $\eta^*(\omega)$ at their respective temperatures. Borg and Pääkkönen [1]

revealed that the new model introduced in this study is four times more precise than the previous models.

12.1 Relations to Molecular Dynamics

Molecular dynamics related to different flows modelled using control theory is discussed. A relaxation-time schema is a good starting point for rough estimations of viscoelasticity. We can approximate relaxation times or generate the entire relaxation spectrum, but if a single relaxation time does not accurately describe the rheology on a macro scale, how would a tube equilibration time τ_e or disengagement time τ_d work better on a molecular scale? According to this consideration, we used the dumb-bell model in a statistical tube.

A further development from a simple linear elastic (Hookean) dumb-bell model is the finitely extensible non-linear elastic (FENE) dumb-bell model [2] that describes three-dimensional flows using the Fokker–Planck equation, yielding a distribution function. Also other micro–macro methods for multiscale simulations of viscoelastic flow using molecular models of kinetic theory have been discussed in the wide-ranging review by Keunings [11]. In our studies, we have converted these micro-level stochastic processes to the statistical micro–macro scale distributions of elastic $P'(\omega)$ and viscous $P''(\omega)$ functions, as is discussed below.

Viscosity and other friction-dependent properties such as self-diffusion can be modelled indirectly using the free-volume theory with the Doolittle formula. Another approach is to use the non-equilibrium Brownian dynamics of Hookean and FENE dumb-bells with internal viscosity, as discussed by Hua and Schieber [12], where internal viscosity is a non-inertial dumb-bell utilizing friction points with Stokes law for drag force and Brownian forces. As a starting formula obtained from the modified Stokes law [13] for the viscosity and restricting force, we use $F'' = 6\pi\xi R\dot{x}$, where $\dot{x}$ is a general rate and ξ is the friction coefficient.

Many studies have investigated relaxed and entangled chains, but few have investigated shear-oriented molecules. We show an alternative physics-based procedure using an oriented chain structure and structural models with the dumb-bell analogy for the chain dynamics in a statistical tube.

We partitioned the complex viscosity formula Eq. (5.8) into dynamic moduli components G' and G'', where the model for viscosity $\eta^*(\omega) = \eta_0^* 10^{-\log\left(\frac{\omega}{\omega_c}\right)(P'(\omega)+P''(\omega))}$. Originally [1] we used natural logarithmic exponent e for these exponent formulas, but to achieve congruence for the P' and P'' constants, we have used the common logarithmic exponent 10 in this book.

For simplicity, we wrote an exponent without relation $-\log\left(\frac{\omega}{\omega_c}\right)$ to set up by Eq. (5.9), a very practical cumulative polymer structure function Eq. (5.12) for $P(\omega)$ with non-logarithmic variable as follows: $P(\omega) = P'(\omega) + P''(\omega)$. Thus elastic polymer structure function $P'(\omega)$ has simple form shown in Eq. (12.1):

$$P'(\omega) = P' \int_{-\infty}^{\log \omega} w'(\log \psi) d \log \psi \qquad (12.1)$$

Elasticity value P' (or actually the inverse value $1/P'$ for elasticity, as shown below) can be explained by molecular dynamics using a simple modified elastic dumb-bell model. We start from the original model [2], where F_c is the force in one dimension exerted by the connector on the particle. The elastic force for a single chain is $F_c = \frac{3kT}{R_0^2}x$. Since the novel mean-square end-to-end distance in equilibrium R_0^2 can lead to complicated treatments close relaxed states, we used an alternative presentation for the spring constant ($K = \frac{3kT}{R_0^2}$) at an oriented steady high flow in the absence of chain entanglements.

Before writing linear relations for Eq. (1) for the elastic connector force $F' = Kx$, the original form of viscoelastic formulas of presented control theory written in simplified form with elastic component $\eta = 10^{-P\prime(\omega)} = 10^{-P\prime}10^{\int()} = \frac{1}{10^{P'}}10^{\int()}$ needs to be considered. Thus, we can write linear relations for elastic force $F' = Kx \propto \frac{\int()}{P'}$, where the integral $\int()$ yields the displacement state x of chains compared with fully relaxed molecules. This integral of course relates also to the amount of entanglement and other chain dynamics. A value of x close to zero indicates a high degree of entanglement, while for a fully oriented state x is 1.

a. Oriented unit segment at high shear drawn with roughly scaled dimensions used for modelling. Possible long chain branching (LCB) increases statistical unit segment length L_s, and increasing temperature expands tube diameter D_s.

b. Behaviours of P' and P'' values and their inverse values are related directly to the elasticity ($1/P$') and viscosity ($1/P''$) of the polymer.

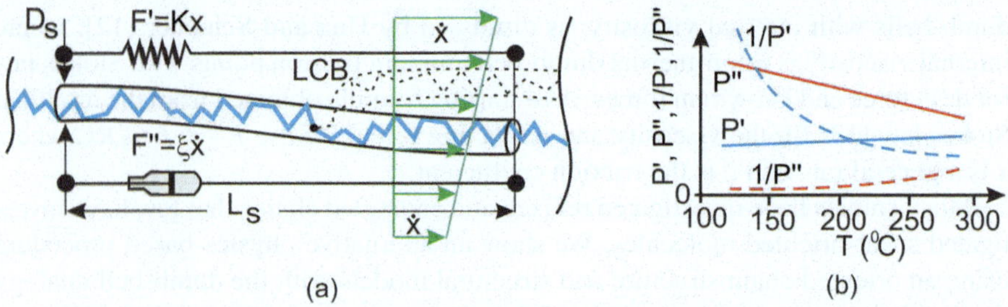

Fig. 12.2 Dimensions of an oriented unit segment in a tube and graph of P' and P'' values at different temperatures. Borg and Pääkkönen [1]

Since different definitions of the "tube diameter" and chain lengths are used in the literature, to avoid confusion we use here the statistical radial tube diameter (D_s) and the statistical unit segment length (L_s), as illustrated in Fig. 12.2a. The ratio L_s/D_s relates to spring constant K, since a larger tube diameter D_s produces a smaller K due to the chains being able to relax more in the radial direction and thereby reduce the stress along the backbone inside the tube. Backbones can have several degrees of rotational freedom in their bonds with different potential energies. If tube diameter D_s is small, backbones have fewer degrees of freedom, which can increase the stress along the tube for the same strain, resulting in a greater spring constant K. If we include the time influence, the relaxation process takes longer for a small tube of smaller diameter at a lower temperature.

We obtain elastic function P' (ω) of Eq. (1) or product ${}^1/_{P'} \int() = kKx$, where scaling factor k includes all the effective molecules of distribution $w(\log M)$ and the integral relates to the respective displacement state x. Thus, we obtain a simple formula for elasticity and inverse elasticity value $1/P'$ using a statistically scaled spring constant $ks = kK$:

$$
{}^1/_{P'} = k_s \frac{L_s}{D_s} \tag{12.2}
$$

We add temperature coefficient a'_T to Eq. (2) to model changes at different temperatures. The values of $k' = 1/k_s$ and a'_T differ between Analytical and Unified Models due to differences in the values of P' and formula structures. Increasing temperature expands the tube diameter (increasing D_s) and respectively decreases the inverse value $1/P'$, producing a corresponding P' increase in Eq. (3). If the polymer has branches, statistical unit segment length L_s increases (or D_s decreases) relatively for a unit backbone, elasticity value P' increases and inverse value $1/P'$ decreases in Eq. (3) during a temperature increase, as obtained in practice for chains of the same chemical type:

$$
P' = k' \frac{D_s}{L_s} + a'_T (T - T_0) \tag{12.3}
$$

For viscosity value P'' we can start from internal viscous restricting force for a statistical unit segment $F'' = \xi\dot{x} \propto P''(\omega)$, where ξ is the internal friction coefficient between chains. A higher temperature increases tube diameter D_s and internal friction coefficient ξ decreases, making it easier for chains to flow at a lower viscous restricting force. We now construct a relation similar to L_s/D_s related to the friction and viscosity forces of the formula $^1/_{P''} f() = k\xi\dot{x}$. Thus, at a higher temperature inverse viscosity value $1/P'$ increases and P'' decreases. We obtain the formula for the relationship between a''_T and viscosity value P'' shown in Eq. (10.4) as

$$P'' = k''\frac{D_s}{L_s} + a''_T(T - T_0) \tag{12.4}$$

The above principle applies to the complex viscosity, but the procedure is also valid for relaxation modulus or shear viscosity. In the Unified Model, P' and P'' have smaller values due to the algebraic structure of functions P' (ω) and P' (ω). P' and P'' values could also influence each other, but this was not investigated. In summary, from the original standpoint of molecular dynamics, the viscoelastic model should use the inverse values for P' and P''; but, as shown below and illustrated in Fig. 12.2b, the presented form is even graphically linear when modelling P' and P'' and is better for the practical modelling of viscoelasticity.

The next task is to find zero viscosity η_0 from molecular dynamics, from which it is possible further to compute zero modulus, as showed earlier. Applying the Rouse-Bueche model for unentangled melt polymers [14] yields

$$\eta_0 = \frac{b^2 M \xi \rho N_A}{36 M_0^2} \tag{12.5}$$

where N_A is Avogadro's number and M_0 is the molecular weight of the monomer (M/N).

Equation (12.5) applies only for unentangled chains, chain topology with the close relaxed state of an entangled molecule is the most complicated, and the possibility of long chain branches and other chain structures add complexity that generates variation in η_0. Measurements made in different studies [15–18] indicate that after reaching plateau modulus G_N^0 at higher frequencies at $\omega > 100$/s, complex modulus $G^*(\omega)$ or viscosity $\eta^*(\omega)$ and η values approach each other for different molecular weights.

Thus, we use critical limit rate ω_∞ and the respective limit complex modulus G^*_∞ or viscosity η^*_∞, where all chains are oriented and unentangled. Applying the ω_∞ to the integral of RED' does not further influence the response to the normalized limit value 1. This critical limit frequency for the example shown in Fig. 12.1 is $\omega_\infty = 10^5$/s, where the M or $w(M)$ RED' no longer has an influence. Thus, we can write Eq. (12.5) for critical limit complex viscosity η^*_∞ as follows:

$$\eta^*_\infty = \frac{L_s^2 \xi \rho N_A}{36 M_0} \tag{12.6}$$

This point is normally outside the measurement range, but does not generate difficulties for software and the principle has already been used for years in practice, with these $\eta^*_\infty(\omega_\infty)$ absolute target points for viscosity found in practice depending only on the used polymer chemical structure and temperature. The MWD can be used to compute zero viscosity η^*_0 and also zero modulus G_0, as shown in our previous studies. Again the above procedure is valid also for shear viscosity η.

12.2 Temperature Dependence Modelled Using Control Theory

We first use control theory to generate absolute values for zero viscosity $\eta^*_0\,(T)$ and as a function of temperature, and then use complete formulas at different frequencies or as a function of time. We use a procedure similar to that used in our previous studies to directly obtain the Analytical formula for zero viscosity by using RED^{T} $w(\log T)$ and temperature value P^{T}:

$$\log \frac{\eta^*_0(T)}{\eta^*_0} = -P^T \int_{-\infty}^{\log T} w(\log \theta)\left(\log \frac{T}{\theta}\right) d \log \theta \tag{12.7}$$

The right-hand side of the formula and RED^{T} relate to the tube diameter D_{s} and to the expansion property. Of course, any frequency can be used for the viscosity; here we use zero states due to their usage in formulas. A simpler Unified formula can be obtained using the characteristic temperature T_{c}, whose optimal value is at the midpoint of the used and modelled temperatures. Characteristic temperature value $P^{\mathrm{T}}_{\mathrm{c}}$ is used as described above:

$$\log \frac{\eta^*_0(T)}{\eta^*_0} = -P^T_c \log \frac{T}{T_c} \int_{-\infty}^{\log T} w(\log \theta) d \log \theta \tag{12.8}$$

RED^{T} $w(\log T)$ can be found by solving the above formula for $w(\log T)$ as we have derived in previous chapters:

$$w(\log T) = -\frac{d}{d \log T} \frac{1}{P^T_c} \frac{\log \frac{\eta^*_0(T)}{\eta^*_0}}{\log \frac{T}{T_c}} \tag{12.9}$$

The model previously developed using control theory (5.3) for complex viscosity $\eta^*(\omega)$ can be written also in short exponent form $\eta^*(\omega) = \eta^*_0 10^{-P(\omega)}$, as can Eq. (12.8) as $\eta^*_0(T) = \eta^*_0 10^{-P^{\mathrm{T}}(T)}$, where $P^{\mathrm{T}}(T) = P^{\mathrm{T}} \int_{-\infty}^{\log T} w(\log \theta)\left(\log \frac{T}{\theta}\right) \mathrm{d} \log \theta$. Thus, we can obtain the complete Analytical model by substituting for complex viscosity as a function of frequency and temperature:

$$\eta^*(\omega, T) = \eta_0^* 10^{-P(\omega)-P^T(T)} \tag{12.10}$$

The respective Unified formula for complex viscosity can be written as

$$\eta^*(\omega, T) = \eta_0^* \log \frac{\omega}{\omega_c} \log \frac{T}{T_c} 10^{-P(\omega)-P^T(T)} \tag{12.11}$$

In a similar manner we can write a temperature-dependent formula for the Unified relaxation modulus:

$$G(t, T) = G_0 \log \frac{t}{t_c} \log \frac{T}{T_c} 10^{-P(t)-P^T(T)} \tag{12.12}$$

For shear viscosity we have Unified formula

$$\eta(\dot{\gamma}, T) = \eta_0 \log \frac{\dot{\gamma}}{\dot{\gamma}_c} \log \frac{T}{T_c} 10^{-P(\dot{\gamma})-P^T(T)} \tag{12.13}$$

The above equations can also be connected to Analytical polymer-structure functions $P(t)$, $P(\omega)$ and $P(\dot{\gamma})$, used instead of analytical temperature function $P^{\mathrm{T}}(T)$, and even used with a mixture of these functions due to the presence of linearity.

12.3 Time–Temperature Superposition

The presented principle can be used to explain TTS and modifications thereof. All the relaxation phenomena involved in $G(t)$ have same temperature dependency, and hence changing the temperature of a measurement will have the same effect on the data as shifting the data horizontally on the log(time) or log(frequency) axis [14]. Dealy and Plazek [19] reports on the practical use of TTS and milestones, which includes a clear description of where TTS holds and fails. We therefore refer to their analysis and use the same nomenclature (e.g. horizontal shift factor a_{T} and vertical shift factor b_{T}). Thus we can write that complex viscosity, which involves both stress and time, requires the application of both shift factors. The master curve is constructed by means of a log–log plot of $\frac{b_{\mathrm{T}}}{a_{\mathrm{T}}}\eta^*(\mathrm{T})$ versus ωa_T.

We can explain TTS in a similar manner to Cox-Merz rules Sect. 5.7 starting from the MWD and using melt calibration by converting viscoelastic data between different deformation modes and temperatures:

$$w(M) = \left|\eta^*(T)\right|_{\eta^*(T_0),P(T),P^T}; \left|\eta(T)\right|_{\eta(T_0),P(T),P^T}; \left|G(T)\right|_{G(T_0),P(T),P^T} \tag{12.14}$$

where $P(T) = P' + P''$ as a function of temperature as indicated in Eqs. (12.2) and (12.3). The same constant (MWD $w(M)$) in Eq. (12.14) is used as the starting and reference point this time only for frequency in Fig. 12.3 and for different REDs and

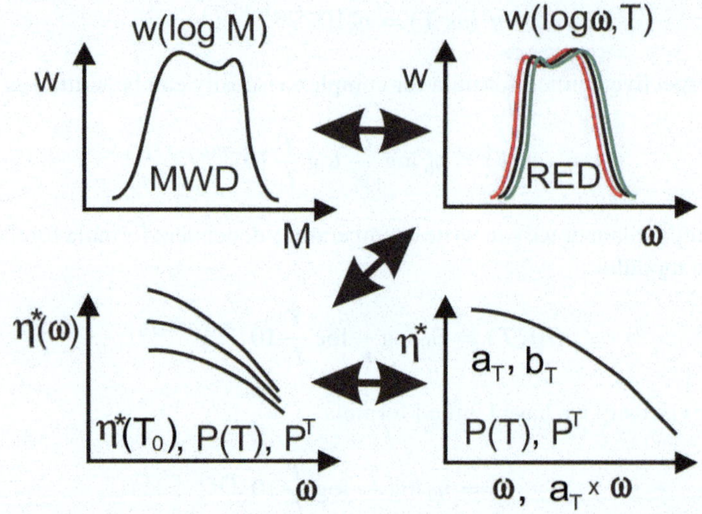

Fig. 12.3 Analyses at different frequencies using the presented method of melt calibration yield the MWD that remains the same for different temperatures. TTS holds if the vertical (b_T) and horizontal (a_T) shift factors give the same viscosity functions as $P(T)$ and P^T. In this case we have a thermorheologically rather simple material, and we can obtain by eye the same master curve (i.e. without visible differences) in the lower-right log–log plot. Borg and Pääkkönen [1]

temperatures. As explained above, RED $w(\log \omega, T)$ is not temperature sensitive, and in practical applications does not exert a significant effect. Since TTS uses linear operators a_T and b_T in a log–log plot and control theory on linear scales, we can write the following approximation for each scale:

$$\eta^*(\omega, T) = \eta_0^* 10^{-P(\omega)-P^T(T)} \cong \frac{b_T}{a_T}\eta^*(T) \quad \text{versus} \quad \omega a_T \tag{12.15}$$

The vertical (b_T) and horizontal (a_T) shift factors have empirical or semi-empirical models as for the Arrhenius and WLF equations, or they can be determined directly from the variation with temperature. Master curves can be used to describe the viscoelasticity of linear polymers over a broad range of times, frequencies or shear rates. Equation (12.15) explains the use of both a single shift factor and both shift factors simultaneously for thermorheologically simple materials, which has many practically applications for TTS. Analysing TTS and Eq. (12.15) from the standpoint of control theory reveals that there can be equivalence if the MWD (and especially RED) has the symmetric form of a normalized Gaussian or similar smooth distribution. However, differences appear with more complex distributions, such as a bimodal one.

In a similar way, blends of different polymers or polymers with LCB can be thermorheologically complex, but in most cases they can still modelled using the control theory principle of wide scales at different temperatures. Moreover, the generated wide master curves violate the presented control theory principle where they join at different temperatures, as obtained in practise when importing master-curve data into the RheoPower software package.

12.4 Experimental Procedure, Test Polymers and Constants

The viscoelasticity and temperature dependence with chain dynamics was modelled. The measured complex viscosity was simulated as a function of frequency and also at a constant frequency as a function of temperature using control theory. The results are compared here with those obtained using Arrhenius and WLF equations. Finally, we discuss more generally used principles, hypotheses, and structures of models and formulas.

Complex viscosity simulations were performed using an Analytical model of control theory on a standard PC with the commercial RheoPower software package, and other simulations were performed using the experimental version of the RheoAnalyzer program of the RheoPower family of software.

The used polymer was Lupolen 1840H LDPE measured at temperatures from 130 °C to 250 °C as in previous studies. The oscillation rheometer data indicated that $Mf = 37{,}800$ g/mol and $Hf = 2.05$. The average molecular weight $M_w = 235{,}000$ g/mol and $M_w/M_n = 14.1$ were used. The used sampling band for the Analytical model was $T = 100{,}000$/s. Measured MWD by GPC/SEC has changed during years as discussed in Sect. 3.8, but may have different MWDs during years, the RED is always constant related to the viscoelastic properties.

The values of P' and P'' obtained from the viscosity fitting procedure are listed in Table 12.1, and no ad hoc constants or values were used. A constant characteristic temperature of $T_c = 190$ °C was used.

12.5 Temperature Dependence of Viscosity by Chain Dynamics

Temperature dependences of viscosity are presented in different ways. As a starting point, complex viscosity η^* as obtained using an Analytical model is shown in Fig. 12.4; we previously presented only simulations obtained using a Unified Model of the same measurements. With Analytical Model the percentage root-mean-square error function *%RMSE* was 100 times higher than that for the respective Unified Model, but was still less than that for any other model of viscosity. Making the sampling band for the Analytical model $T = 1{,}000$/s instead of $T = 100{,}000$/s (as used here) would reduce *%RMSE* by more than 50%, but we wanted to use as wide a sampling band as it is reasonable in order to maximize the applicability of the Analytical model.

Table 12.1 Main characteristics of all computations. $1/P'$ and $1/P''$ values are not linearly related

T (°)[a]	$1/P'$ [b]	$1/P''$ [b]	$P' + P''$ [c]	R [c]	η_0^* [d]	%RMSE[e]
130	6.56	0.08	13.31	1,122,018	566,867	0.2480
140	1.64	0.09	12.10	1,122,018	357,525	0.4415
150	1.53	0.09	11.84	1,000,000	239,109	0.3548
160	0.93	0.10	10.79	891,251	137,602	0.4790
170	0.76	0.11	10.13	891,251	93,747	0.5107
180	0.60	0.13	9.40	794,328	54,081	0.5455
190	0.51	0.15	8.48	794,328	35,967	0.6087
200	0.47	0.16	8.51	707,946	23,885	0.5714
210	0.43	0.17	8.10	630,957	17,523	0.5759
220	0.39	0.20	7.52	630,957	12,056	0.5345
230	0.37	0.22	7.17	562,341	9,128	0.5436
240	0.34	0.27	6.60	562,341	6,283	0.6194
250	0.32	0.29	6.62	501,187	4,647	0.5493

[a] Temperature
[b] Obtained elasticity P' and P'' viscosity values
[c] Sum of P' and P''
[d] Distance R set for the ratio of effective distribution ranges
[e] Obtained percentage root-mean-square error function

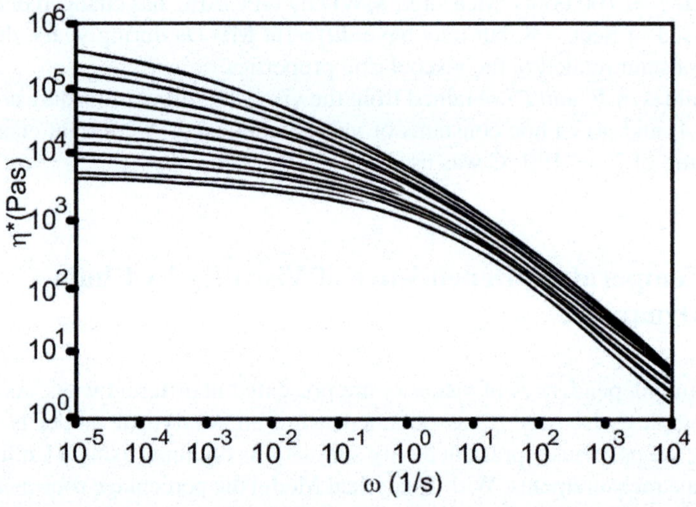

Fig. 12.4 Complex viscosity η^* modelled using Analytical models for Lupolen 1840H LDPE at temperatures ranging from 130 °C (top curve) to 250 °C (bottom curve) in 10 °C increments. In Fig. 12.5 we modelled the Unified complex viscosity for the same measurements. Borg and Pääkkönen [1]

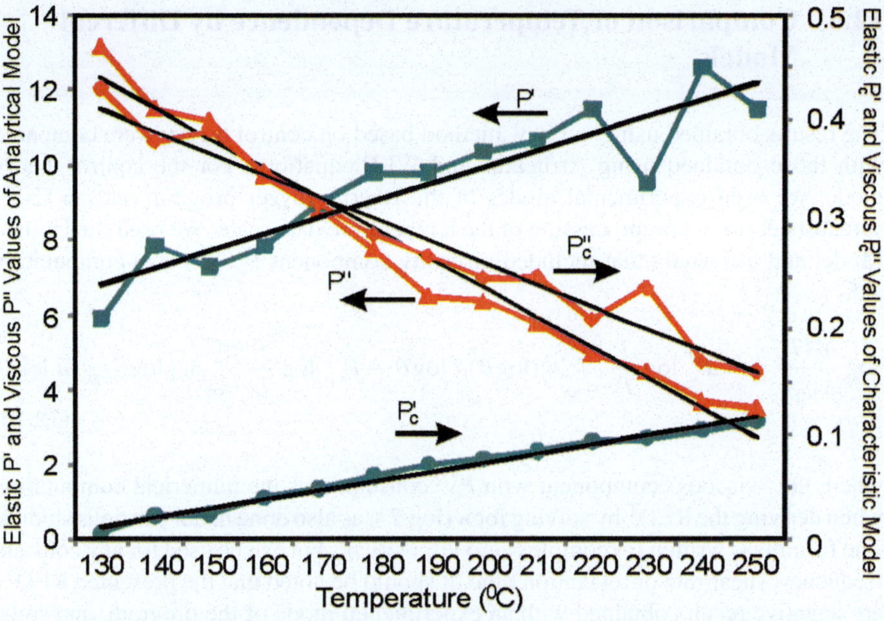

Fig. 12.5 Elasticity P' and viscous P'' values obtained by modelling viscosity η^* with an Analytical model (left scale) and a Unified Model (right scale) marked $P'_{\ c}$ and $P''_{\ c}$ respectively at different temperatures. Small measurement errors were found for the model at 230°C. Black straight lines are regression fits. Elastic and viscous regression fits have same directions, only absolute values differ on the left and right scales. Borg and Pääkkönen [1]

We used the obtained elasticity P' and P'' viscosity values to study chain dynamics. Figure 12.5 indicate that P' and P'' behave very linearly as a function of temperature and according to formulas for chain dynamics (Eqs. (12.3) and (12.4)).

We selected a tube diameter $D_s = 3.6$ nm [20] and a statistical segment length L_s $= 50$ nm as it is in range 8 nm–60 nm [15, 19], but accurate absolute values were not essential in this study. Using our values we obtained a friction coefficient of ξ $= 10.8 \cdot 10^{-10}$ kg/s at 150 °C using Eq. 12.6. The available literature suggests that ξ $= 2.3 \cdot 10^{-10}$ kg/s and density 785 kg/m^3 [20], which yields a critical limit complex viscosity of $\eta^*_\infty = 1.26$ Pa s, which is the same as that modelled at $\omega = 100{,}000$/s in Fig. 3. From this value it is easy to obtain the complete viscosity flow curve for all types of polyethylene if the MWD, P' and P'' are known.

Using simple linear modelling we obtained $k' = 10.8$ (elastic scaling factor) and $a'_T = 0.0016$/°C (temperature coefficient) with Eq. (12.3), and $k'' = 9.1$ (viscous scaling factor) and $a''_T = -0.0021$/°C (temperature coefficient) with Eq. (12.4). Similar linear behaviour was found with the Unified Model, with the P' and P'' values only being lower: $k' = 57$, $a'_T = 0.0245$/°C, $k'' = 239$ and $a''_T = -0.080$/°C, as shown in Fig. 12.5. This linear behaviour confirms the validity of the linear model obtained using control theory for chain dynamics also as a function of temperature.

12.6 Comparison of Temperature Dependence by Different Models

The results obtained using our new method based on control theory were compared with those obtained using Arrhenius and WLF equations. For the control-theory model we used experimental modes of the RheoAnalyzer program and an Excel spreadsheet for the preprocessing of the temperature data scale. We used the Unified Model and a formula that included viscosity component $P_c^{''T}(T)$ as a computation aid:

$$\log \frac{\eta_0^*(T)}{\eta_0^*} = -P_c^T \log \frac{T}{T_c} \int_{-\infty}^{\log T} w(\log \theta) d \log \theta - P_c^{''T} \log \frac{T}{T_c} \int_{-\infty}^{\log T} w\left(\log {}^{\theta}/_{R}\right) d \log \theta \tag{12.16}$$

where the "viscous" component with $P_c^{''T}$ constant aids the numerical computation when deriving the RED^{T} by solving for $w(\log T)$, as also done in our previous studies. The formula is written for complex zero viscosity η_0^*, but can be used for any constant frequency, shear rate or relaxation time. It should be noted that the presented RED^{T}s are tentative results obtained with an experimental mode of the program, and some coding work is needed to obtain more accurate results.

For the Arrhenius equation we used the standard formula shown in Eq. (12.17), and for the WLF equation we used Eq. (12.18), where we use vertical shift factor b_{T} instead of the normally used a_{T}:

$$b_T = \exp\left[\frac{E_a}{R}\left(\frac{1}{T} - \frac{1}{T_0}\right)\right] \tag{12.17}$$

$$b_T = \frac{-c_1^0 (T - T_0)}{c_2^0 + T - T_0} \tag{12.18}$$

Data for zero viscosity $\eta_0^*(T)$ were modelled using an Analytical model and taking the data shown in Fig. 12.4. Note that the data for the Arrhenius and WLF equations were generated directly from these equations, whereas those for the control-theory model required preprocessing in an Excel spreadsheet generating some errors to feed data. Applying a standard best-fit routine yielded $E_a = 73.5$ kJ/mol for the activation energy, and empirical fitting parameters yielded $c_1^0 = 6.95$ and $c_2^0 = 309$ K for zero viscosities. R in the WLF equation is the universal gas constant. We get results shown in Fig. 12.6.

Viscosity $\eta^*(T)$ was also modelled at $\omega = 1$/s as a function of temperature using control theory and the Arrhenius and WLF equations. This yielded $E_{\mathrm{a}} = 21.0$ kJ/mol, $c_1^0 = 1.2$ and $c_2^0 = 220$ K. Modelling using control theory was found to be four times more accurate by *%RMSE* compared to the Arrhenius equations and six times better than with the WLF equation, although some absolute measurement errors were found that were possibly due to the edge effects of polymer melt sample. An

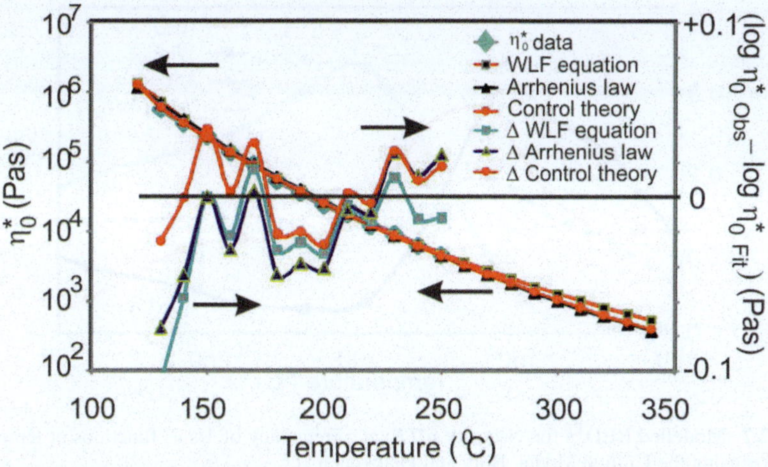

Fig. 12.6 Zero viscosity $\eta_0^*(T)$ data obtained from the analytical model shown in Fig. 12.3 as a function of frequency, and compared as a function of temperature with the results obtained using the Arrhenius and WLF equations, and control theory. $\eta_0^*(T)$ modelled using control theory is four times more accurate. Borg and Pääkkönen [1]

oscillation rheometer is known to be a relatively accurate device for frequency-sweep measurements, but differences can appear in the absolute values obtained for different sample sets of measurement runs. Since we only modelled zero relaxation modulus G_0 generated by zero viscosity, there was sense in making comparisons for G_0.

12.7 Rheologically Effective Distribution RED^{T} for Temperature

We computed the RED^{T} function using the Arrhenius and WLF equations, and control theory, and first solved Eq. (12.16) for distribution $w(\log T)$ using a similar principle to that described previously in the Chapter 9:

$$w(\log T) = -\frac{d}{d\log T}\frac{1}{P_c^T}\left(\frac{\log\frac{\eta_0^*(T)}{\eta_0^*}}{\log\frac{T}{T_c}} + P_c^{''T}\int_{-\infty}^{\log T} w\left(\log\frac{\theta}{R}\right) d\log\theta\right) \qquad (12.19)$$

Now we can compute limited RED^{T}s, since these distributions tend to widen below 100 °C and above 350 °C, which are not physically possible for a melt. Thus, we obtained RED^{T}s from curves generated using Arrhenius and WLF equations, and control theory at a frequency of 1/s. The distributions were derived directly from Arrhenius and WLF equations for viscosity $\eta^*(T)$. Viscosity data $\eta^*(T, 1/\mathrm{s})$ were

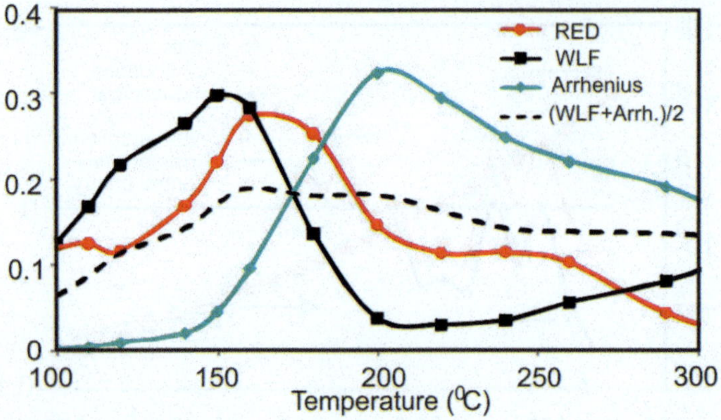

Fig. 12.7 Modelled RED^{T}s for viscosity $\eta^*(T)$ at a frequency of 1/s as functions of temperature obtained using the Unified Model. Borg and Pääkkönen [1]

taken from Fig. 12.4, developed by an Analytical model, and the results are shown in Fig. 12.7.

12.8 RED^{T}s as Functions of Temperature and Frequency

Obtaining the three-dimensional RED^{T}s as functions of temperature and frequency can provide more useful information, and are generally known as hyphenated techniques.

The generated three-dimensional plot of RED^T curves in Fig. 12.8 shows that temperature influences the distributions, with another bump being evident around 270 °C. The physical meaning of the bump is not yet clear, but may have something to do with minor additional chain freedoms and tube diameter.

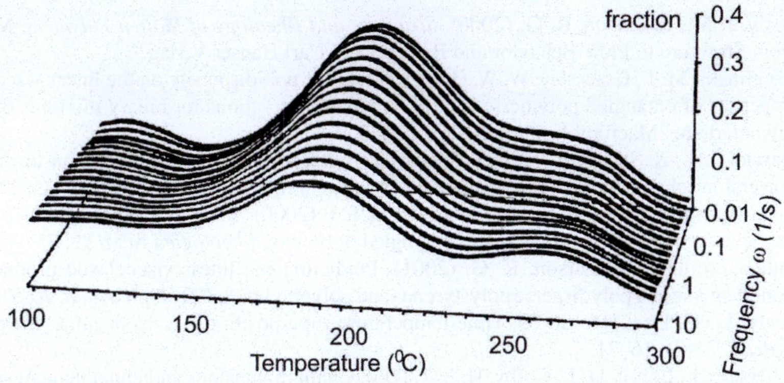

Fig. 12.8 Computed three-dimensional family of RED_T $w(\log T, \omega)$ curves for various temperatures and frequencies. This type of chart can give more structural information, and is generally known as hyphenated techniques. We can obtain that at higher frequencies fewer changes to the tube diameter are needed at different temperatures. Borg and Pääkkönen [1]

References

1. Borg, T., Pääkkönen, E. J. (2010). Linear viscoelastic models. part IV. From molecular dynamics to temperature and viscoelastic relations using control theory. Journal of Non-Newtonian Fluid Mechanics, 165, 24–31.
2. Bird, R. B., Curtiss, C. F., Armstrong, R. C., Hassanger, O. (1987). *Dynamics of polymeric liquids, Vol. 2. Kinetic Theory*, 2nd Ed., New York: Wiley. New York 1987.
3. Van Gurp, M., Palmen, J. (1998). Time-temperature superposition for polymeric blends. *Rheol Bulletin, 67*, 5–8.
4. Wood-Adams, P., & Costeux, S. (2001). Thermorheological effects of long-chain branching in entangled polymer melts. *Macromol, 34*, 6281–6290.
5. Graessley, W. W., Raju, V. R. (1984). Some rheological properties of solutions and blends of hydrogenerated polybutadiene. *Journal of Polymer Science: Polymer Symposia, 71*, 77–93.
6. Lohse, J., Milner, S. T., Fetters, L. J., Xenidou, M., Hadjichristidis, N., Mendelson, R. A., Garcia-Franco, C. A., & Lyon, M. K. (2002). Well-defined, model long chain brached polyethylene. *Melt rheological behavior, Macromolecules, 35*, 3066–3075.
7. Macaubas, P. H. P., & Demarquette, N. R. (2002). Time-temperature superposition principle applicability for blends formed of immiscible polymers. *Polymer Engineering and Science, 42*, 1509–1519.
8. Giamberini, M., Ambrogi, V., Cerruti, P., & Carfagna, C. (2006). Viscoelasticity of main chain liquid crystalline elastomers. *Polymer, 47*, 4490–4496.
9. Kagarise, C., Koelling, K. W., Wang, Y., & Bechtel, S. E. (2008). A unified model for polystyrene–nanorod and polystyrene–nanoplatelet melt composites. *Rheol Acta, 47*, 1061–1076.
10. Kim, J., Kim, D. H., & Son, Y. (2008). Abnormal rheological behavior of linear low density polyethylene melts at high shear rate. *Polymer Bulletin, 60*, 821–828.
11. Keunings, R. (2004). Micro-macro methods for the multiscale simulation of viscoelastic flow using molecular models of kinetic theory. In: Binding, D. M., Walters, K. (Eds.). Rheology Reviews 2004, British Society of Rheology, 67–98.
12. Hua, C. C., & Schieber, J. D. (1995). Nonequilibrium brownian dynamics of hookean and FENE dumbbells with internal viscosity. *Journal of Non-Newtonian Fluid Mechanics, 56*, 307–332.
13. Batchelor, G. K. (1967). *An Introduction to Fluid Mechanics*. Cambridge, UK: Cambridge University Press.

14. Dealy, J. M., & Larson, R. G. (2006). *Structure and Rheology of Molten Polymers*. Munich: From Structure to Flow Behavior and Back Again. Carl Hanser Verlag.
15. Struglinski, M. J., Graessley, W. W. (1985). Effects of polydispersity on the linear viscoelastic properties of entangled polymers. 1. Experimental observations for binary mixtures of linear polybutadiene. Macromolecules 18, 2630–2643.
16. Mavridis, H., & Shroff, R. (1993). Appraisal of amolecular weight distribution-to-rheology conversion scheme for linear polyethylenes. *Journal of Applied Polymer Science, 49*, 299–318.
17. Pattamaprom, C., Larson, R. G., & Van Dyke, T. J. (2000). Quantitative predictions of linear viscoelastic rheological properties of entangled polymers. *Rheologica Acta, 39*, 517–531.
18. Pattamaprom, C., & Larson, R. G. (2001). Predicting the linear viscoelastic properties of monodisperse and polydisperse polystyrenes and polyethylenes. *Rheologica Acta, 40*, 516–532.
19. Dealy, J., & Plazek, D. (2009). Time-temperature superposition – A users guide. *Rheologica Bulletin, 78*(2), 16–31.
20. Fetters, L. J., Lohse, D. J., Colby, H. R. (2006). Chain dimensions and entanglement spacing. In: Mark, E. J. (Ed.), Physical properties of polymers handbook, 2nd ed., Springer, New York.

Chapter 13
Computation Cases

Some results, originating from MWD and RED functions, have been previously presented in discussions involving various types of polymer flow. Most calculations can be performed using simple tools like calculators or spreadsheets, without the need for complex computations. However, detecting MWD requires a real computer and specialized software. The results in this chapter represent a compilation of data gathered from other chapters, supplemented with new computations.

13.1 Used Viscosity Data

The complex viscosity data used is shown in Fig. 13.1.

13.2 Utilized Software

The RheoPower software package [2] comprises two separate software programs: (1) the RheoAnalyzer mode, which is used to find the MWD from viscosity or relaxation modulus curves; and (2) the RheoDeveloper mode, which is used to find modelled curves from structure. Both programs give simultaneous results over broad ranges of viscosity, relaxation modulus, and spectrum fits.

T. Borg, *Control Theory in Rheology*, https://doi.org/10.1007/978-3-031-88248-7_13

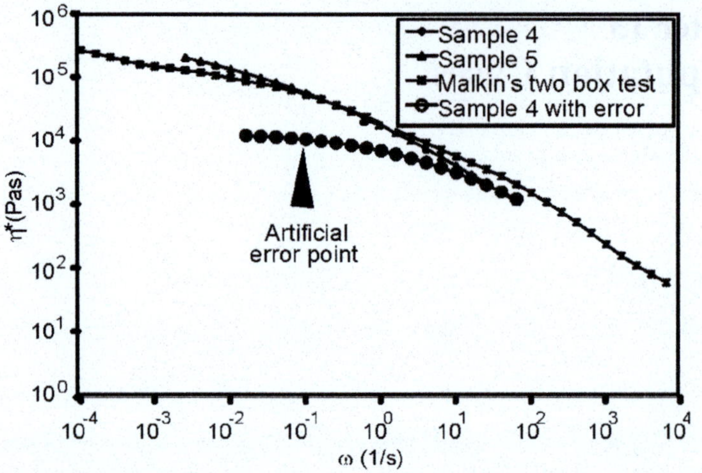

Fig. 13.1 Complex viscosity of Borealis HDPE Sample 4, Sample 5, fictional Malkin's two-box test and Sample 4 with artificial error point 1% viscosity error at a frequency of 0.1/s. we generated an artificial 1% viscosity error at a frequency of 0.1/s, by changing the originally measured η^* from 8798 to 8886 Pas. Borg and Pääkkönen [1]

13.3 Short Computation History and Set of Equations

The development of the software for our programs has taken several years. We initially found Gleissle's [3] viscometric mirror relation ($\dot{\gamma} = 1/t$) very helpful, and the second generation of the software package was developed based on differential equations and rheological phenomena. The third generation of the software is compiled in fluent binary form, and can complete the computation in minutes when running on a standard PC. In contrast, computations using the first versions of the software required hours or even days to get MWD result. The modelling of viscoelastic functions from structure is always done in seconds.

13.4 Some Results to Get MWD

Only two polymer-dependent constants are used. For the HDPE sample, Mf was 52,900 g/mol and the default set of $R = 10^6$ was set for the distance for the Rouse shear range distribution. The computed MWD curve for Borealis HDPE Sample 4 is shown in Fig. 13.2. For the LDPE sample, $Mf = 37{,}800$ g/mol at 150°C was used. The relation scale exponent for PE, $Hf = 2.05$, was held.

13.5 Evaluation of Sensitiveness of Method

The two-box test by Malkin [4] was used to simulate transformations from the MWD to the viscosity flow curve, and vice versa. The RheoDeveloper software was first used to draw two cubic boxes, simulating bimodal MWDs. The computed viscosity curve was saved in the databases and used by the RheoAnalyzer software. The results are shown in Fig. 13.4, which indicates the generation of minimal noise that only slightly distorts the MWD boxes. The source of the noise is known and coming from tuned derivation, but the software is for practical computations. Moreover, *Mw* and *MwR* are close to their original values Fig. 13.3.

13.6 Simulation of Sensitivity to Data Error

A simulation test was performed to determine the effect of a small (i.e., 1%) error in the viscosity data point on the MWD results. Taking the measured viscosity data of Borealis Sample 4 used in Fig. 13.1, we generated an artificial 1% viscosity error at a frequency of 0.1/s, by changing the originally measured η from 8,798 to 8,886 Pas. On a logarithmic scale, this changed the data point from 3.9444 to 3.9487, which was impossible to detect visually on the flow curve. The results are shown in Fig. 13.5.

Figure 13.5 shows that the computed MWD exhibits significant distortion on the HMW side, with %RMSE increasing from 0.0013 to 0.0018%. It must also be borne in mind that even more severe errors are caused by measuring-system deviations such as slipping, poor mixing of compounds, or poor relaxation of the sample. The error evaluation shows that it is impossible to detect the MWD from a manually copied viscosity curve. In this method, moderate absolute errors can be present in the viscosity, but the relative errors must be minimal.

13.7 Evaluation of Error

The calculation of the MWD is extremely sensitive to errors in the recorded data. The influences of error sources and the precision of records on the estimation are summarized in Fig. 13.6.

De Vries et al. [5] used %RMSE values of viscosity for the well-defined metallocene-based molecular structures of polyolefin elastomers and plastomers. Their shear-rate range was $0.1/\text{s} < \omega < 10{,}000/\text{s}$. The values for %RMSE were in the range 0.6–1.3% for the model of Bersted et al. and 0.8–1.5% for the Cross model; in their (1996) neural net model, %RMSE was in the range 1.2–2.5%.

A simple test revealed that it was possible to copy viscosity values manually to a precision level of log 0.01, resulting in an %RMSE on the order of 0.5%. The values

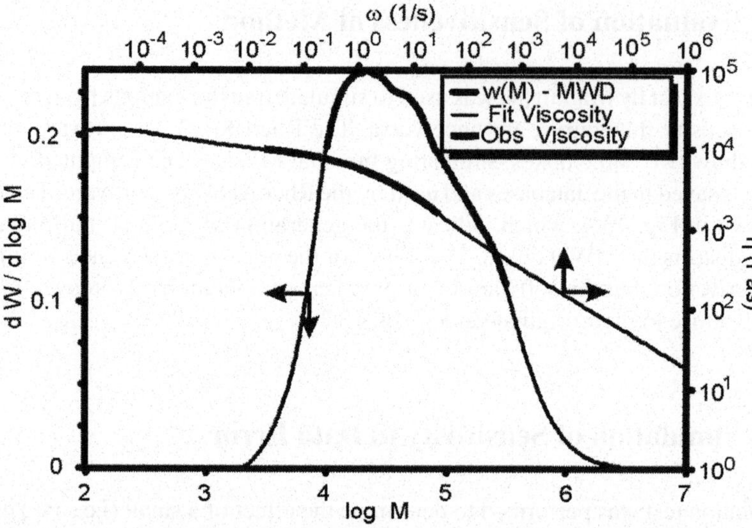

Fig. 13.2 MWD curve for Borealis HDPE Sample 4. The thicker segment of the MWD curve has been computed directly from viscosity data by derivation. The viscosity curves are also shown for measured (thick) and computed (thin) segments. Only two polymer-dependent constants are used. For the HDPE sample, *Mf* was 52,900 g/mol and the default set of $R = 10^6$ was set for the distance for the Rouse shear range distribution. The respective Values obtained from GPC measurements for Sample 4 were $\overline{M}$w = 132,000 and $MwR = 4.1$, which indicates that the error was at the level of numerical errors (with %RMSE = 0.0014%). Borg and Pääkkönen [1]

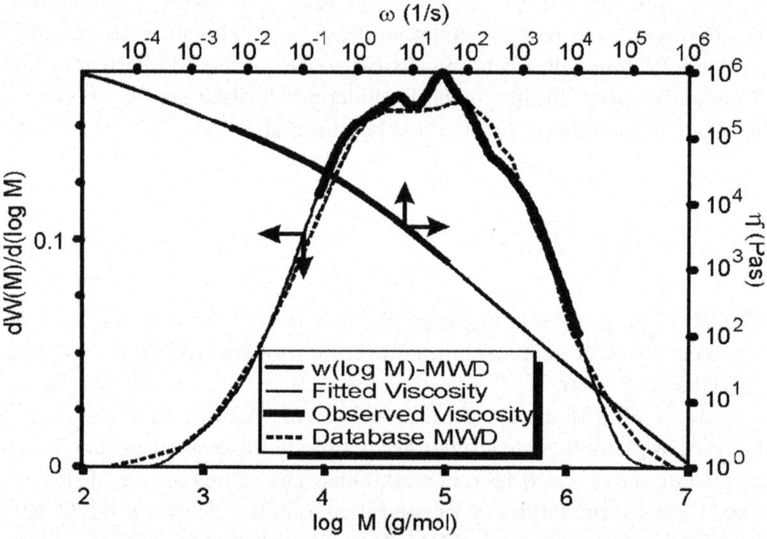

Fig. 13.3 MWD curve for Borealis HDPE Sample 5. The received GPC data are labelled as "Database MWD" on the chart. The viscosity curves are also shown for measured (thick) and computed (thin) segments. Borg and Pääkkönen [1]

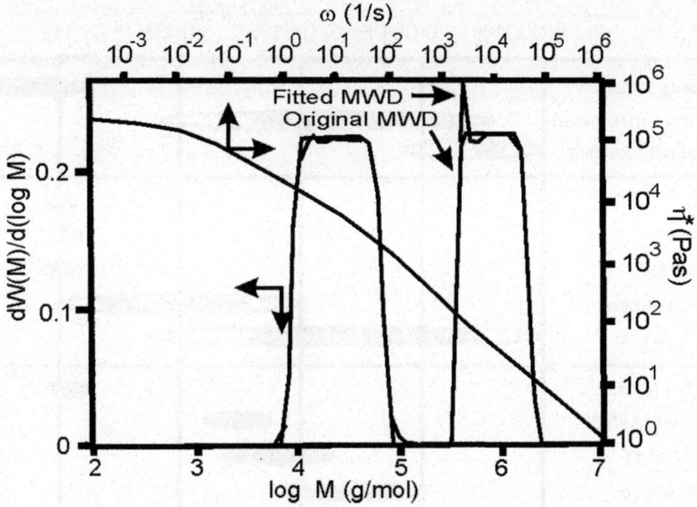

Fig. 13.4 Fictional Malkin two-box test. The viscosity curve was generated by the RheoDeveloper (MWD-Visco, i.e., from MWD to the properties) software, starting from the data labelled "Original MWD". The viscosity scale is arbitrary, but the relative changes are in agreement with the theory. Using the generated viscosity curve, the MWD results shown here were generated again using RheoAnalyzer (Visco-MWD). The recalculated fitted MWD is very close to the original MWD. Borg and Pääkkönen [1]

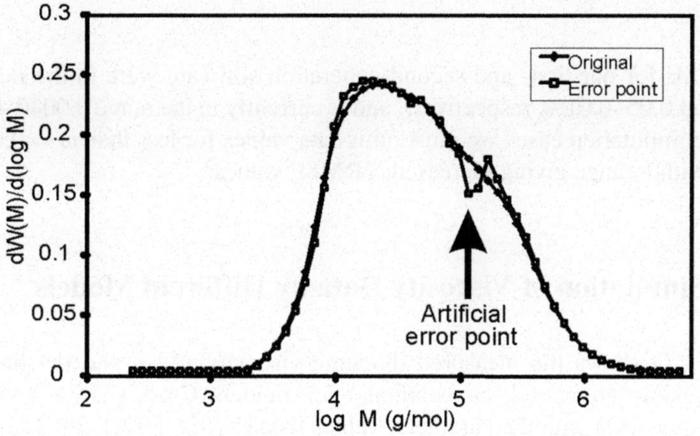

Fig. 13.5 Simulation test for the MWD shows how a small (1%) error in a single viscosity data point can distort the MWD curve of Sample 4 (shown in Fig. 13.2). An artificial frequency error of 0.1/s was added to the viscosity. A small (1%) data error resulted in nearly twenty times (approximately 18%) the local error in the MWD curve. Borg and Pääkkönen [1]

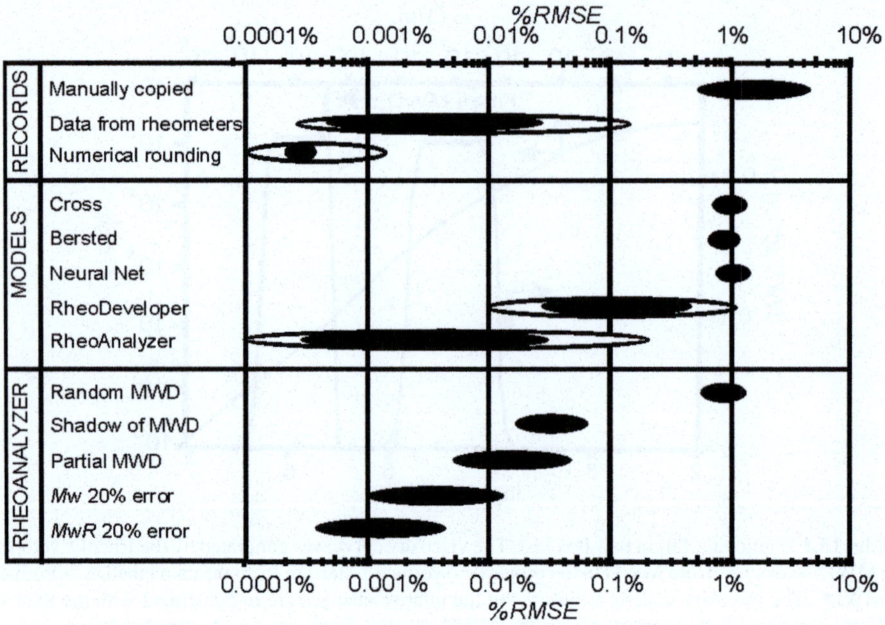

Fig. 13.6 Typical %RMSE errors for different known models compared to the computation levels of RheoAnalyzer. The fitting procedure makes our method more than 100-fold more accurate than the other models. Borg and Pääkkönen [1]

of %RMSE for our first- and second-generation software were in the ranges 0.5–0.05% and 0.05–0.01%, respectively, and is currently in the range 0.0040–0.0003%. In most computation cases we must omit data values for less than ω = 0.001/s and records with a range giving increased %RMSE values.

13.8 Simulation of Viscosity Data by Different Models

Figure 13.7a shows the measured dynamic viscosity $\eta^*(\omega)$ and the flow curves obtained using various well-known rheological models (Cross, Carreau and Carreau-Yasuda equations) with the parameter values listed in Table 13.1 [6]. The best variable fits to the measured data were obtained using a spreadsheet, while with the RheoPower package, zero viscosity η_0^* was assumed at a frequency of 10^{-6} 1/s, where the Analytical model shows a η_0^* value that is too high. Pure control theory includes all size chains of the MWD and the effects of the complete RED at very low frequencies are too strong, while the modified characteristic model derived using control theory produces a more realistic fit.

The quality of the fit between the observed η_{obs}^* and predicted η_{fit}^* viscosity values was quantified using a modified least-square procedure that yielded the percentage

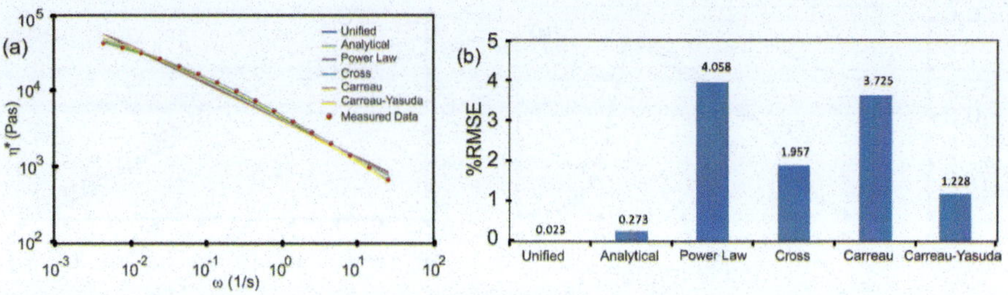

Fig. 13.7 Dynamic viscosity of Lupolen 1840H LDPE at 150 °C. **a** Measured flow curves and those modelled using various well-known rheological models. **b** The quality of the fit between the observed η^*_{obs} and predicted η^*_{fit} viscosity values for the different models as determined from the percentage root-mean-square error (% RMSE) Discussed in Sect. 13.8. Borg and Pääkkönen [6]

Table 13.1 Parameter values used in different models for simulations of viscosity data

Polymer and flow	η^*_0	λ	m	k	n	a	p
Analytical	167,283						
Characteristic	73,800						
Power law				4.025	0.519		
Cross	100,200	288	0.546				
Carreau	72,400	330					0.248
Carreau-Yasuda	173,000	88			0.31	0.3	

root-meansquare error (% RMSE). Least-square procedures are widely used in numerical computations, and the differences obtained in the present study are shown in Fig. 13.7b.

The used viscosity LDPE data are curved and difficult to fit to some models, which is why the %RMSE values of other models are more than twofold higher than normal. The characteristic and Analytical models actually included the form of data for the RED function, which partly explains why the observed error is small. Moreover, the characteristic viscosity fits might still be accurate when extrapolated outside of the measurement range over a wide scale of 10^{-6}/s < ω < 10^{6}/s since they are computed from the true MWD/RED.

13.9 What-If Analyses

Other properties and the influence of changes in the MWD on the behaviour can be simulated using what-if analyses presented in Fig. 13.8. This enables the background of principle, where is possible to compute to the both directions from MWD or viscoelastic measurements.

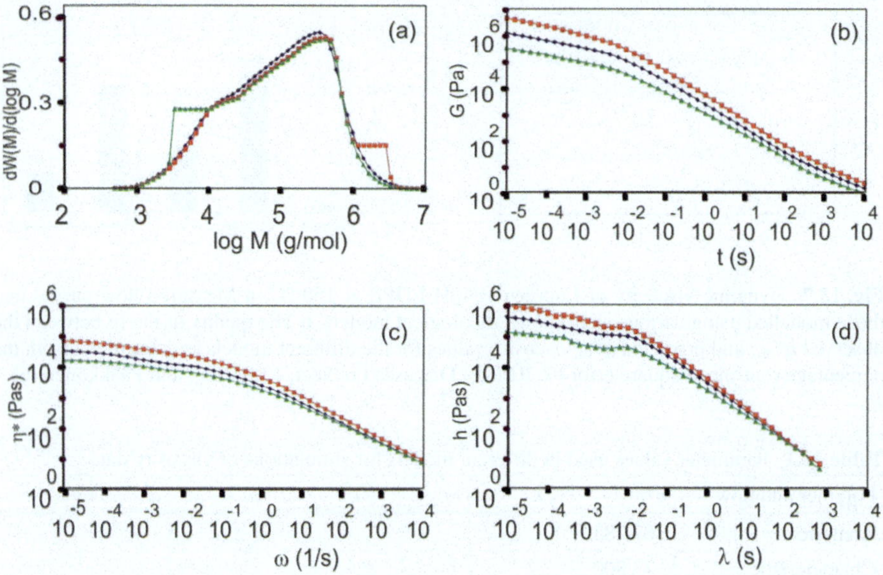

Fig. 13.8 What-if analyses for Lupolen 1840H LDPE at 190°C reveals large changes in modelled viscosity, relaxation modulus, and spectrum. **a** Small differences are evident in the cubic shape of the MWD. Original values of *Mw* and polydispersity *MwR* are 230,000 g/mol and 11, respectively. $Mw = 202{,}000$ g/mol and $MwR = 11.7$ for the left-side cubic bump, and $Mw = 296{,}000$ g/mol and $MwR = 11.0$ for the right-side bump. **b** Relaxation modulus $G(t)$. The maximal relaxation time is not set for the spectrum maximal relaxation time, and the maximal relaxation time is not used so as to demonstrate the cross-point of the spectrum in Fig. 13.8d. **c** Modelled shear viscosity starting from MWD (modelled here as complex viscosity η^*). **d** Modelled relaxation spectrum $h(\lambda)$. Borg and Pääkkönen [6]

13.9.1 More General Summaries

More general summaries of the models and formulas are now provided for different flows described here. The used hypotheses are derived from fundamental principles of molecular dynamics and structure by statistical procedures with explicit formulas.

At first is showed that logarithmic convolution derived from control theory is a linear operation, since it is commutative, associative and distributive. All the formulas presented are linear and explicit, but also uses standard and implicit numerical methods in practise to derive RED $w(\omega)_{\text{obs}}$. Similarly, in the complete computation system of orientation simulations using finite element modelling is an implicit procedure.

A new form of the Boltzmann superposition principle used and the parallel arrangement of forces led to the novel idea of using time-rate separability for transient viscosity and for the relaxation function. The physical model for chains starts from well-known fundamentals and the novel (but still logically presented) melt calibration, with the influences of chains discussed in more detail. All other hypotheses presented and the use of linear and explicit formulas, control theory, and the derivation of RED from data form seamless causal correlations with obtained measurements. In other words, correlations with viscoelastic measurements involved three variables being conducted (the RED function and elasticity P' and viscous P'' values), where the RED can be converted even to the bimodal MWD or more complicated forms. This manuscript describes how P' and P'' values exhibit linear behaviour relating to chain dynamics and dimensions. In short, we obtained stable statistical correlations and viscoelastic fits simultaneously.

Ill-posed inverse problems are discussed, which in this case tend towards a well-posed state. Most attempts in the field have led to integral equations, which are known to be ill-posed.

Inverse problems usually do not fulfil Hadamard's postulates of well-posedness; namely, they might not have a solution in the strict sense, or the solutions might not be unique and/or might not depend continuously on the data. Since $w(\omega)_{\mathrm{obs}}$ is computed directly by derivation from measured data using a Unified Model and is converted into relative fractions of the MWD $w(M)$, this segment curve fulfils the definition of well-posedness.

In contrast, the Analytical model used to obtain the MWD is highly ill-posed.

Some information from the difference viscosity curve $\Delta\eta^* = \eta^*_{obs} - \eta^*_{fit}$, and from the direction of %RMSE convergence to develop $w(\omega)_{\mathrm{fit}}$ side curves. Thus, these out-of-range regions are still mildly ill-posed.

The wide η^*_{fit} and $G(t)$ fulfil the well-posedness definition, even though the curves are based on a form of extrapolation (actually in the linear model) that works in all directions. If the model is highly consistent with measurements –within the range of numerical round-offs – it can be expected to provide accurate estimations of new data outside the measured data range.

In contrast, although the temperature-dependency model correlates empirically well with measurements according to mathematical treatments similar to those presented above, chain models and the origin of certain behaviours are partially affected by temperature-dependent viscoelasticity (Fig. 13.9).

13.9.2 Discrepancy of Analytical Model

One novel finding was that not all molecules impact simultaneously to the flow properties. During developing of the analytical model for complex viscosity we took in the use sampling band T for the data collection. In the oscillating rheometers also real shear rates for molecules varies a lot and frequency ω is a kind virtual integral of all shar rates at ω rate. This introduced the development of the Unified Model. The discrepancy of Analytical model during shear viscosity found recently were more dramatically as shown in the Fig. 13.8.

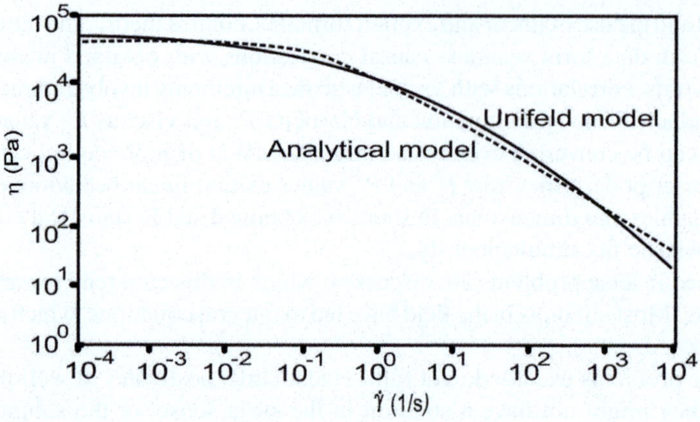

Fig. 13.9 Comparison of shear viscosities developed by analytical and unified models are shown. Unified viscosity is practically the same as according to measurements. The difference to analytical results were still greater as with complex viscosity data models

References

1. Borg, T., Pääkkönen, E. J. (2009). Linear viscoelastic models. Part II. Recovery of the molecular weight distribution using viscosity data. Journal Non-Newtonian Fluid Mech 156, 129–138.
2. https://www.tomcoat.com
3. Gleissle, W. (1980). Two simple time-shear rate relations combining viscosity and first normal stress coefficient in the linear and non-linear flow range In G. Astarita., G. Marrucci & L. Nicolais (Eds.) Rheology, vol 2. Plenum Press, p 457
4. Malkin, A. Y. (1990). Some inverse problems in rheology leading to integral equations. *Rheologica Acta, 29*, 512–518.
5. De Vries, S. A., Kale, L. T., Oswald, T., Karjala, T. P. (1996). Molecular structure–Linear viscoelastic properties modeling for constrained geometry catalyst polyolefin resins. In A. Aït-Kadi., J. Dealy., D. James & M. Williams (Eds.) on Rheology, Quebec, Canada 18–23 August 1996.
6. Borg, T., & Pääkkönen, E. J. (2015). Linear viscoelastic model for different flows based on control theory. *Appl Rheol, 25*, 1–10.

Chapter 14
Overview of the Unified Model

This chapter provides an overview of the Unified Model with MWD analyses.[1] The Unified is designed for analysing different types of polymer flow and MWD and LCBs (long Chain Branches). More about detecting MWD by Unified compared GPC/SEC methods is also discussed.[2, 3]

The variance in MWD results across different laboratories measured by GPC/SEC has also been discussed, underscoring the challenges in developing MWD analysis models based on viscoelastic measurements. Over the decades, we have observed developments in GPC/SEC analyses for PE, but throughout that time, the RED function of polymer flow and behaviour has remained constant. With current polymer types or newly developed batches, the situation may be similar, with the RED of a polymer batch remaining consistent.

14.1 Introduction

The analysis procedure is introduced using the terminology of liquid gel permeation and size-exclusion chromatography methods for detecting the respective chromatogram or elution curve called the rheologically effective distribution (RED), and this can be further converted into the molecular weight distribution. The RED is the direct linear fingerprint of the molecular and dynamic structure and of viscoelastic properties such as shear and complex viscosities and relaxation modulus. The Unified Model is based on linear control theory and melt calibration without employing

[1] Presented at the International Conference on Polyolefin Characterization (ICPC), September 21–24, 2014, in Valencia, Spain.

[2] T. Borg, E. J. Pääkkönen, Polymer structures and flows by new linear viscoelastic principle (2013) non-published.

[3] T. Borg, E. J. Pääkkönen, A new computational rheological method for determining polymer structure from viscosity measurements (2015) non-published.

T. Borg, *Control Theory in Rheology*, https://doi.org/10.1007/978-3-031-88248-7_14

relaxation-time spectum concepts. The RED is formulated starting from the polymer micro scale for statistical tube dimensions up to the macro-scale viscoelastic flows, and is a versatile tool for molecular characterization and viscoelasticity modelling. Most rheological methods are sensitive for long-chain branches (LCB), but the novel presented procedure provides more information about both LCB and their corresponding distribution.

The analysis method presented herein can use data from different types of polymer flow. However, but this study is focused on the data obtained from an oscillating rheometer to determine the molecular-weight distribution (MWD) and other structural information. The presented method is more versatile for different types of devices, but this study focused on the dynamic rheometer and complex viscosity. The measuring procedure used by a rheometer adds mechanical energy to the sample in order to generate a shear viscosity or stress as a response, from which the rheologically effective distribution (RED) is computed.

Different aspects of this wide subject are first discussed to provide some background to the development of the measuring practices and the formulas used. The influence of the measuring practices used in oscillatory rheometer on the determined MWD is presented, including by addressing whether the excitation frequency sweeps should be ascending or descending. The internal smoothing effect of gel permeation chromatography (GPC) and size-exclusion chromatography (SEC) procedures is simulated in some demonstrative examples. The computation results for long-chain branches (LCB) and their corresponding distribution (LCBD) are also presented in Sect. 11.13 "Simulated MW-LCB distributions" as Sect. 14.5 "Long-Chain Branches Distribution (LCBD)."

14.2 Problems Encountered While Developing the Method

A short review about detecting the MWD from viscoelasticity measurements. The large fundamental differences in the underlying mechanics and techniques between analytical GPC and the more-or-less strict mathematical and rheological methods for detecting MWDs have resulted in unsuccessful outcomes and frustrations over decades for many accomplished rheologists due to the poor reference data. The initial work performed during the 1970s and 1980s used viscosity measurements as data sources for detecting the MWD [1–16], followed a decade later by the use of the relaxation modulus or spectrum [16–28]. Many related studies have also been reported in the 21st century [29–35], but recently very few of them have used polyethylene (PE) [36–41] or polypropylene (PP) [42] samples. No significant difficulties were encountered when applying the various procedures to polystyrene (PS) and other amorphous polymers, but the use of GPC/SEC by different laboratories for determining the MWD of PE in particular has yielded misleading results for developers of rheological models as discussed in Sect. 3.8. This could be one reason why recent studies determining the MWD from rheological data have mainly involved non-PE materials. Nevertheless, there has been some progress in the embarrassment of obtaining the MWD of PE. One source of error associated with using GPC/SEC analyses to determine the MWD of PE is presented in Sect. 11.6. Differences in the results obtained by individual laboratories are first discussed below.

14.3 Interlaboratory Accuracy of GPC/SEC Methods

The problems associated with using GPC/SEC to analyse certain polymers are universally known, and this prompted several round-robin tests during the 1990s by several research groups worldwide [43, 44] and by IUPAC work groups [45–47]. These analyses based on round-robin tests have involved the same samples being tested by dozens of different international laboratories, polymerization plants and universities. One overall finding is that the differences in the obtained MWDs were minimal for the easy measurable and soluble PS, in contrast to the results obtained for PE samples.

Reference [49]. That work group found differences in the results from different laboratories of up to threefold for the average *Mw* and *MwR* values, as indicated in the chart. The identification number of each laboratory and the used calibration method are listed along the bottom of Fig. 14.1.

Round-robin tests for SEC performed in 2010 [48] showed that the repeatability within a laboratory was low (~1%), with large standard deviations. However, comparisons between different laboratories revealed a reproducibility of up to 30% (based also on standard deviations). The wide variations are assumed to be caused by the use of different SEC systems, even when identical calibration standards are employed.

Another round-robin project by IUPAC SEC/GPC performed in 2010 [49] found that the average *Mw* for low-density PE (LDPE) varies roughly from 100,000 to 300,000 g/mol, as shown in Fig. 12.1. The polydispersity (PDI) *MwR* values have ranged from below 6 to over 12, while the absolute values have differed by two- to threefold.

The general practice of showing average *Mw* values on a g/mol scale is misleading, at least for PE values—the kg/mol unit would be more appropriate. The observed summarized GPC/SEC results exhibit relatively good intralaboratory repeatability and very poor apparent interlaboratory reproducibility. The great number of unknown differences in the GPC/SEC results for the MWD of PE from different sources has impeded the development of rheological methods for detecting MWDs.

No doubt developments in accuracy have occurred a lot as shown MWD results in Fig. 3.4, but differences can still be between laboratories.

14.4 Ascending or Descending Frequency Sweeps?

Using frequency sweeps when measuring the dynamic viscosity causes deviations in the detected MWD. The technical report of the recent IUPAC round-robin tests of polyolefin melts using rotational rheometers [50] revealed differences of approximately 40% in the reported zero-shear-rate viscosity. There are many reasons why descending frequency-sweep tests are more convenient for measurements using oscillation rheometers, but these may result in valuable information about the possible LCB contents and especially the LCBD being overlooked. The difference in the measured reversible shear thickening between using ascending and descending frequency sweeps is very obvious for dispersed PMMA [51] and even more so for colloidal suspensions [52]. In contrast, differences for polymer melts are not so obvious, and the phenomena can even be invisible unless additional computations are performed. Schematic Fig. 14.2 illustrates the situation in detecting LCBD. Obtained results with polymer melts and reasons are discussed in the experimental Sect. 11.13.

Relaxation of long LCB chains may need hours. Complete relaxation may take even over night in nitrogen gas, if possible to carry.

In principle, the steady shear viscosity data discussed in Chap. 4 should be more effective for detecting MWD, as measurements encompass a broader range of RED/MWD data. However, the capillary rheometer has limitations due to pressure and entrance effects. The Sandwich rheometer (SSR) would be ideal from the standpoint of Unified analyses, but it may face practical and instrumental limitations. One of the purposes of this book is to elucidate the potential of SSR for accurate MWD and LCB analyses.

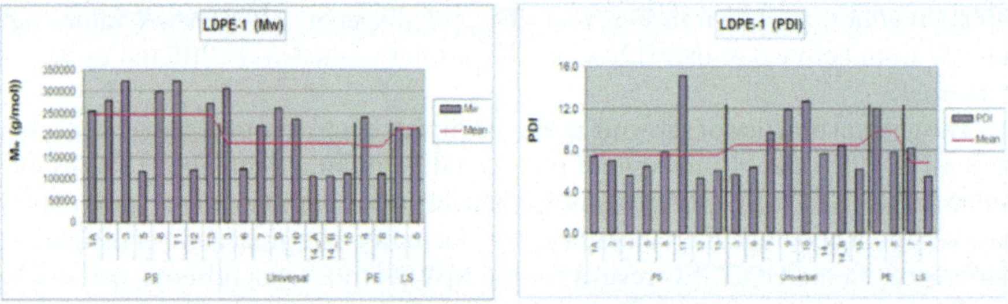

Fig. 14.1 Results for the *Mw* and *MwR* values of LDPE in charts copied from the IUPAC SEC/SEC Round Robin Project Report (2010)

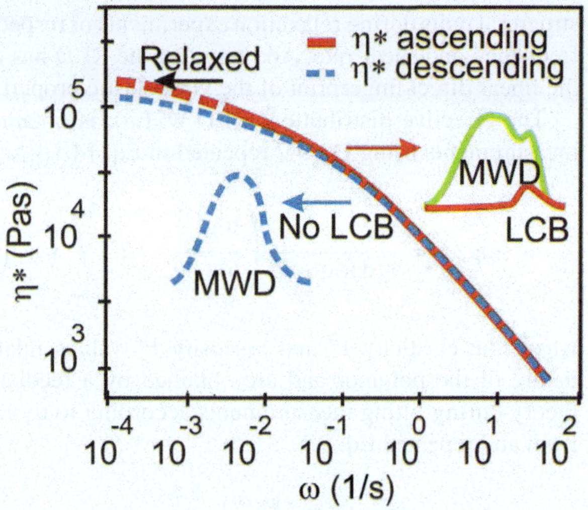

Fig. 14.2 Schematic illustrating magnified differences for dynamic viscosity data. The red lines indicate data measured using an ascending frequency sweep, which yields well-relaxed LCB at low frequencies and the location of the LCBD. The blue lines indicate data measured using a descending frequency sweep, which yields similar MWD data with GPC results

14.5 Using Unified Model to Analyse Polymers

The background to the new model differs from those underlying previously used theories in rheology and viscoelasticity. Polymer flows are modelled and micro structures are detected more accurately than by the earlier control theory, without using relaxation times, in contrast to traditional Maxwell-type models. The Maxwell model is mathematically valid and is descriptive for introducing viscoelasticity. Unfortunately, the formulation principle means that this model cannot accurately account for the various non-linear viscoelastic behaviours of real polymers influenced by long molecules, since the orientation and amount of chain entanglements vary with the flow state. Although the spectrum itself is an elegant tool, it has limitations when considering the flows of complex microstructures. Modern control theory is used widely in many technologies, including digital computers, aircraft design and processing industries, but it has not been applied previously to viscoelasticity. A brief overview of the background and the use of control theory is available elsewhere. Detailed explanations about the background principles and formulas have previously been presented for different flows as shear in Chap. 4, the complex viscosity in Chap. 5, relaxation modulus in Chap. 6, transition in Chap. 10 and elongation viscosities in Chap. 11; we repeat here only the information needed for practical polymer analyses based on oscillation rheometer data.

We introduce the RED, and how to convert this into the MWD by melt calibration; that is, the relation between time or frequency and the molecular weight fractions. This procedure has similarities with the widely used universal calibration with dissolved polymers. The RED function can be regarded as a GPC trace with time-exclusion chromatography or SEC elugram function as a function of time. The largest molecules exit first from the GPC/SEC column, in a similar way to these molecules initially having greater relaxation effects with many entanglements and structural units during relaxation experiments or respectively orientating at the lowest frequency and shear rate. Additionally, the RED has the important feature of being the linear direct fingerprint of the viscoelastic properties of polymers.

The effective distribution, RED $w'(\omega)$, is obtained from rotational rheometer measurements using Eq. 9.2 repeated in Eq. 14.1.

$$w'(\log\omega) = -\frac{\mathrm{d}}{\mathrm{d}\log\omega}\frac{1}{P'}\left(\frac{\log\frac{\eta^*}{\eta_0^*}}{\log\frac{\omega}{\omega_c}} + P''\int_{-\infty}^{\log\omega}\left(w''\left(\log\frac{\psi}{R}\right)\right)\mathrm{d}\log\psi\right), \tag{14.1}$$

where the elasticity P' and viscosity P'' values relate directly to the viscoelastic nature of the polymer and are obtained by a feedback procedure used in control theory during fitting measurements according to dynamic viscosity $\eta^*(\omega)$ obtained from and repeated Eq. 9.1.

$$\log\frac{\eta^*(\omega)}{\eta_0^*} = -\log\frac{\omega}{\omega_c}\int_{-\infty}^{\log\omega}\left(P'w'(\log\omega) + P''w''\left(\log\frac{\psi}{R}\right)\right)\mathrm{d}\log\psi \tag{14.2}$$

This loop between Eqs. (14.1) and (14.2) is performed around thousand times during a single computation to obtain the best feedback fits to the RED and viscosity function according to measured data. In summary, we obtain linear relations that are accurate to within the numerical rounding errors between the RED and dynamic viscosity $\eta^*(\omega)$. This procedure does not use measured elastic and viscous components directly, but instead these are extracted by computation and can be partitioned back into modelled dynamic modulus components G' and G'' as shown in Sect. 5. 4. This approach avoid some distortions of the distribution between the measured storage and loss components.

14.5.1 Using Unified Model for Steady Shear Viscosity

Steady shear viscosity data, discussed in Chap. 4, meet all the criteria for obtaining good and accurate RED results in principle. This approach scans the entire range of polymer sizes in the MWD with a narrow spectrum, unlike other flow types that involve mixed shear rates and, consequently, mixed molecular sizes. Unfortunately, mechanical limitations may restrict the ability to conduct these measurements.

14.6 Conversion from the RED to the MWD by Melt Calibration

In a manner somewhat analogous to GPC/SEC methods involving the chromatogram being converted into the MWD, we convert the RED into the MWD by melt calibration. When an effective distribution function (RED′) w' (log ω) is found to be the best fit for the viscosity, we can convert w' (log ω) into a function of the molecular weight, MWD $w(\log M)$, using a conversion factor between scale Hf and the polymer structural value Mf. The background of the melt calibration for frequency has been presented previously in the Sect. 5.5.

We have developed from molecular dynamics a simple relation for the melt calibration, $M(\omega)$, as a function of frequency, where the value of Mf is M at $\omega_1 = 1/$ s:

$$M = Mf\left(\frac{\omega}{\omega_1}\right)^{-\frac{1}{Hf}} \tag{14.3}$$

For polymer measurements, the Hf conversion factor is usually close to 4 (with the exception of older PE data, whose values were $2 < Hf < 3$), and Mf takes values between 10^2 and 10^6 g/mol. Equation (14.3) gives the linear calibration line for a logarithmic scale, which has worked well so far for the relation. If a different form of curved function is found to be a better alternative, this can easily be substituted instead, as is done with GPC/SEC systems.

14.7 Comparison with Liquid Chromatography Systems

Equations (14.1)–(14.3) represent a new rheological analysis method for detecting the MWD and other polymer structures. It is analogous to liquid chromatography methods [53–55] such as GPC [56–58], which is also referred to as SEC, and high-performance liquid chromatography (HPLC) [59]. A general overview of these systems is presented in Fig. 14.3. The rheological method provides wide fits for viscoelastic flows, the relaxation modulus and the spectrum as additional results. Elasticity P' and viscosity P'' values directly relate to the viscoelastic nature of a polymer and the relative dimensions of the statistical tube of the molecules, as discussed previously in Sect. 12.2. Additionally, experimental data can be obtained at different temperatures.

The elastic component RED (or $w(\log \omega)$) is computationally extracted from the measured viscosity data, and this is the main contributor to changes in the response to applied forces, and is linearly related to the viscoelastic properties in a similar way to the component RED″ (or w'' (log ω/R)) at lower frequencies. The function of the RED can be regarded as a trace with time-exclusion chromatography (i.e. GPC)

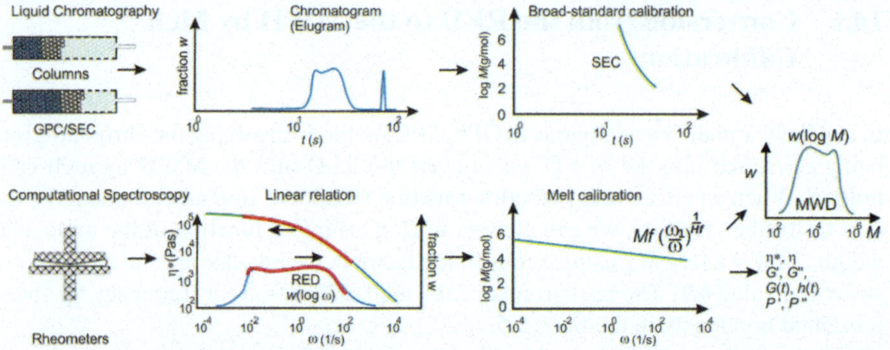

Fig. 14.3 Schematic procedures of liquid chromatography (upper panels) and the respective rheological methods (lower panels) to obtain the same MWD. The measured dynamic viscosity and the accurately computed RED curve segments are indicated by thicker lines in the middle lower panel. Wider fits were obtained by least-square procedures and a best-fit routine

or an elugram with SEC. As the respective eluent, this elastic component varies with changes in the active oriented and disentangled chains and their segments.

Some differences between the methods are as follows: Chromatography separates concrete physical particles [60] as molecules. The rheological method is a technique using data not only from rotational rheometer, but also capillary rheometers or other similar devices for measuring viscosity or stress in different states or as a function of time. The new method is mathematically valid and linear in control theory. The presented procedure does not use directly measured storage and loss components and computing forms an essential part of the procedure. Storage and loss components are computed afterwards from viscosity, complex modulus or relaxation data and results are rather close with the measured storage and loss moduli. We could therefore even refer to the method as "computational rheological spectroscopy".

14.8 Long-Chain Branches Distribution (LCBD)

We now discuss a new procedure to obtain real distributions for the LCBD with partially known MWDs starting from the RED, based on a simple distribution subtraction procedure and melt calibration. Real LCBDs have very rarely been presented with the MWD in the literature.

A nuclear magnetic resonance (NMR) method is used to detect LCB, but it is known that NMR can be used to detect side-chain branches with up to eight carbon atoms only [61]. Many alternative ways have been developed to obtain a single numerical value [62–67] including by utilizing the Zimm-Stockmayer equation [68]. Several rheological methods are suitable for identifying LCB [69, 70], such as van Gurp-Palmen [71, 72] and Cole-Cole [73] plots, flow activation energy [74], thermorheology [75], the dynamic modulus [67], and relaxation times [76–78], but all suffer from the problem of not being able to convert the obtained results into absolute numerical values.

Applying a stochastic approach to topological LCBD models has yielded the multidimensional distributed molecular properties (e.g. joint MWD–LCBD) with the number of LCB [79–81].

We employ a different procedure using a more realistic one-dimensional LCBD with the total number of inner backbones. The left panel in Fig. 14.4a shows the terminology [82] used, with the primary chain as the backbone, directly linked chains as free arms, and the next connected chains as inner backbone segments. Of course the LCBD in Fig. 14.4b can be subsequently fractionalized into the bivariate MWD–LCBD if needed. Main distribution $w_{MAIN}(M)$ is detected rheologically in earlier Chapters, which was used predominantly in text. The final results are shown using the sum of RED or MWD to obtain the normalized sum value of 1; that is, $\int w_{MAIN} + \int w_{LCB} = 1$.

Since GPC/SEC measurements cannot be used to detect LCB, we separate the RED into two distributions: MWD $w_{MAIN}(M)$ and LCBD $w_{LCB}(M)$. We first convert the RED into the respective $w(M)$ distribution, as shown in Fig. 12.4b. The normalized MWD, obtained over a narrower data range from Eq. (14.4) or by GPC/SEC, is used to obtain the distribution marked w_{GPC}. LCBD w_{LCB} is found by subtraction from the true MWD $w(M)$ gained from the RED:

$$w_{LCB}(M) = w(M) - kw_{GPC}(M), \tag{14.4}$$

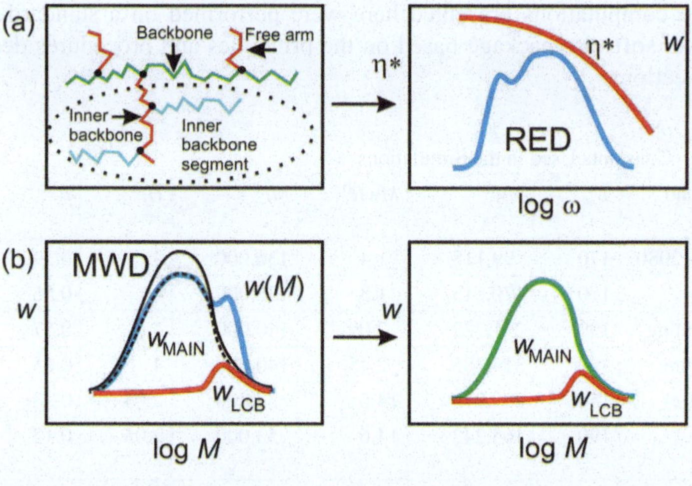

Fig. 14.4 Schematic diagrams explaining how LCBD $w_{LCB}(M)$ is detected. **a** Molecule chain has a backbone, free arms, and inner backbone segments. Viscoelasticity measurements were used to obtain the accurate linear RED here with a second bump. **b** From GPC/SEC or by using the narrow data range (or viscoelastic data measured during an ascending frequency sweep) to obtain $w(M)$, we determined the normalized MWD or $w_{MAIN}(M)$. We obtain the LCBD or $w_{LCB}(M)$ from the difference $w(M) - w_{MAIN}(M)$, where $w_{MAIN}(M)$ is the base MWD without LCB

where k is found easily during by fitting between the distributions.

We can renew the formula for the LCB—expressed as the number of LCB per 1,000 carbon atoms (1000C) as presented in previous studies [67, 70]—using the above results for LCBD as follows:

$$\frac{\text{LCB}}{1000}\text{C} = 1000M_\text{b}\sum\frac{w_{\text{LCB}}(M_\text{i})}{M_\text{i}}, \tag{14.5}$$

where w_{LCB} is the weight fraction of the LCBD, M_i is the molecular weight M of each fraction, and M_b is the molar mass of a backbone element ($M_\text{b} = 14$ for PE). Complete weight fraction w_i of the MWD and the average number of branches per molecule were originally used in the formula [67]. However, now this value consists not only of branches on the backbone but also of inner backbones, as shown in Fig. 12.4a. A major disadvantage of these types of formulas is that the average *Mw* of backbones (and now branches) strongly influence the obtained values.

14.9 Variables and REDs

For melt calibration, polymer structural value *Mf* and conversion factor *Hf* between scales are presented only for comparison to obtain the MWD. The main characteristics of the samples are listed in Table 14.1. The smoothing levels used and the respective obtained Mw and MwR values are shown in Table 14.2.

All the computations presented here were performed on a standard PC using RheoPower software package based on the principles and procedures described in the next sections.

Table 14.1 Constants Used in the Simulations

Polymer and flow type	T^a	Mw^b	MwR^b	Mf^c	Hf^c	P'^d	P''^d
PS gstern_2080	170	99,118	1.4	136,000	4	0.39	0.08
PS gstern	170	170,615	1.3	136,000	4	0.56	0.16
PS descending	190	291,722	2.09	140,000	4	0.57	0.25
PS ascending	190	314,385	2.45	140,000	4	0.65	0.38
LDPE $\eta^*(\omega)$	150	239,973	14.6	37,800	2.05	0.28	0.45
HDPE MWD_{GPC}	190	165,242	14.6	53,000	2.05	0.17	0.43
HDPE $w(M)$	190	320,983	25.3	53,000	2.05	0.19	0.99

[a] Measured temperature in degrees Celsius
[b] Computed average *Mw* in g/mol and PDI index *Mw*/*Mn*
[c] Used structural value *Mf* and conversion factor *Hf*
[d] Elasticity *P*′ and viscosity *P*″ values

Table 14.2 The smoothing levels used and the respective obtained *Mw* and *MwR* values

MWD	*Mw*[a]	*MwR*[b]	log a[c]	log μ[d]	log σ[e]
Two-box as it is	192,104	39.1	0	0	0
Two-box as PS	192,502	39.1	0.02	0.2	0.3
Two-box as PE	230,434	44.1	0.28	0.2	0.3
LDPE 1840 H as it is	200,514	12.4	0	0	0
LDPE 1840 H as PS	200,929	12.4	0.02	0.2	0.3
LDPE 1840 H as PE	240,522	14.0	0.28	0.2	0.3

[a] Computed average *Mw* in g/mol
[b] Computed PDI index *Mw*/*Mn*
[c] Sampling band value *a* for smoothing level
[d] Mean value of normal (Gaussian) distribution
[e] Deviation value of normal (Gaussian) distribution

14.10 The Investigated Mixtures

Chromatograms or REDs are shown for PS mixtures and their respective MWDs. Figure 14.5 shows bimodal results also for REDs, which are additional to the previously reported MWDs for the same experiments shown Sect. 4.3. Frequency-sweep measurements according to our suggested routines as presented above would improve the fit quality. Small measured viscosity errors generate visible roughness in the RED, but this relation is linearly accurate according to Eqs. (14.1) and (14.2).

Figure 14.5 indicates that the zero viscosity differed by tenfold between the two mixtures. Smoothing occurs during GPC/SEC measurements, and MWDs measured

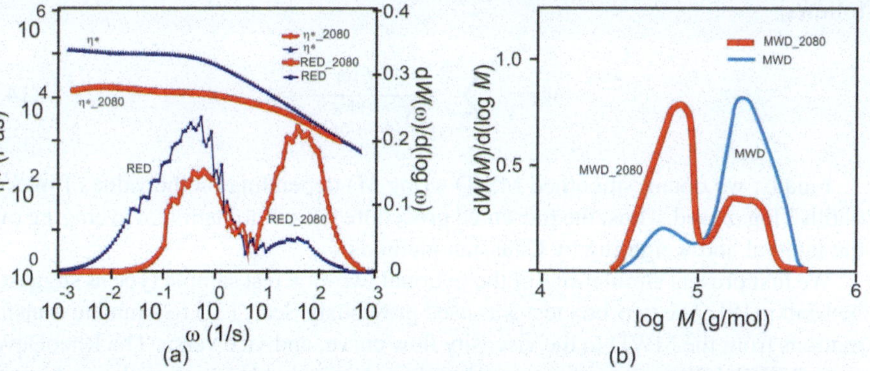

Fig. 14.5 REDs, MWDs, and viscosities for bimodal PS mixtures. PS benchmark test data for “gstern_2080” (thick and dashed blue lines), representing PS mixtures from 20 wt.% (177,000 g/mol PS) to 80 wt.% (60,000 g/mol PS), and for “gstern” (red lines), representing inverse mixtures (ratio of 80:20). **a** REDs derived from viscosity measurements and accurate wider modelled dynamic viscosities. **b** MWDs converted from REDs are practically same as those measured by the GPC/SEC procedure

by GPC/SEC are not shown in Fig. 14.5 for the sake of clarity since the developed and smoothed MWDs were similar to the GPC results. Moreover, the descriptive data for molecular characterization are elasticity P' and viscosity P'' constants, which are not discussed here.

Our analyses indicated that GPC/SEC procedures exert smoothing effects that depend on the polymer. This was revealed by performing numerous comparisons of the computed rheological data MWDs and the GPC/SEC results. It is therefore necessary to generate a smoothing function in the software to obtain results similar to those obtained with liquid chromatography instruments. The degree of smoothing was close to zero with PS but considerable with PE polymers, and his effect greatly influences MwR and the average Mw. We considered this effect using two simulations: (i) assuming a fictional MWD and (ii) performing a true computation from viscosity data.

After testing several mathematical models, we found that modified weighted-average smoothing was the best for simulating the phenomena that occurred internally in the columns or the complete procedure. The simulation model starts from a normal (Gaussian) distribution $f(\log M)$ on the logarithmic scale as follows:

$$f(\log M) = \frac{1}{\sigma\sqrt{2\pi}} \mathrm{e}^{\frac{-(\log M - \mu)^2}{2\sigma^2}}. \tag{14.6}$$

This is a normal standard distribution with the usual variables and mean μ and standard deviation σ, and only difference is the use of the logarithmic scale. In the next step we used the principle of weighted averages of functions. This means we have to use weighted values only on the interval $[w_{\mathrm{i}} - \log a, w_{\mathrm{i}} + \log a]$, and the complete width of the sampling band is 2 log a, as shown in Fig. 14.6. We thereby obtain the formula for each weight fraction w_i from respective Gaussian value f_{i} as follows:

$$w_i = \sum_{i=-\log a}^{i=+\log a} \frac{f_{\mathrm{i}} w_{\mathrm{i}}}{\sum_{i=-\log a}^{i=+\log a} w_{\mathrm{i}}}. \tag{14.7}$$

Finally, we obtain smoothed MWD $w(\log M)$ depending on the value of interval width 2 log a used. Thus, the presented procedure is a combination of averaging over the interval and weighting by Gaussian methods.

We first present simulations of the fictional two-box test sample type as suggested by Malkin [8]. The two-box test was used previously Sect. 13.5 to simulate transformations from the MWD to the viscosity flow curve, and vice versa. The RheoDeveloper (MWD-VIS) software was used to draw two cubic boxes, simulating bimodal MWDs. The smoothing results at different levels are shown in Fig. 12.6.

Figure 14.7 shows computed results from dynamic viscosity data for LDPE 1840 H measured at 150 °C by RheoAnalyzer (VIS-MWD) at different smoothing levels. Due to the graph shape being close to a Gaussian distribution, the effects for PDI are less radical than those described above, but they are still meaningful. Errors visible

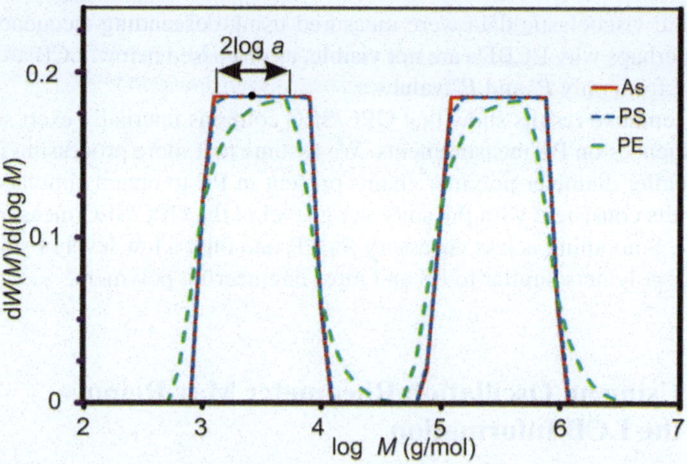

Fig. 14.6 Simulations of different smoothing levels using RheoDeveloper software: "as it is" (As), PS and PE, related to GPC/SEC results. Used levels by the sampling band value a for smoothing levels are original as it is, used with PS and highest for PE, where the average Mw and *MwR* increase by 20% and 13%, respectively. Sampling band 2 log a is illustrated for the data point marked by a dot and PE polymer

in the non-smoothed "as it is" form come from the measured data and effects of numerical rounding.

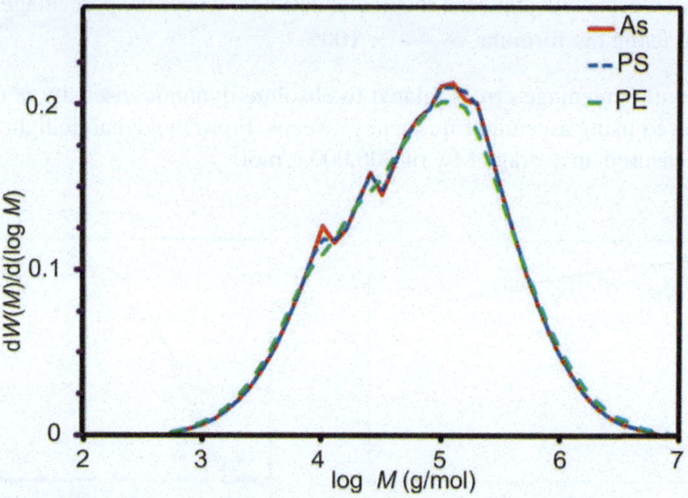

Fig. 14.7 Simulating different smoothing levels of GPC/SEC. The graphs look similar, but the average Mw and MwR actually increased by 20% and 13%, respectively

Dynamic viscoelastic data were measured using descending frequency sweeps, which is perhaps why LCBDs are not visible, as discussed below. LCB are therefore obtainable from only P' and P'' values.

These tentative results show that GPC/SEC columns internally exert some functional influences on PE measurements. We assume that more processing is required for the smaller diameter polymer chains present in PE to obtain simulated computation results consistent with the smoothing level in the GPC/SEC measurements of the MWD. Smoothing is less necessary for PP, and only a low level of smoothing is needed for polymers similar to PS and most engineering polymers.

14.11 Using an Oscillation Rheometer May Remove the LCB Information

There were no visually obvious differences when the viscoelastic data of polymer melts were measured by oscillation rheometers using an ascending or descending frequency sweep, but sensitive software did reveal differences, which also become visible in the computed MWD and RED functions. One major problem encountered when developing the new method was the very high sensitivity of the procedure to small relative errors in the viscosity data. However, the absolute values of dynamic viscosity differences between different frequency-sweep tests produced fewer problems than did tiny local relative distortions in the same run. Figure 14.8 shows the results obtained with the same PS sample measured using ascending and descending frequency sweeps with standard rheometer settings, where the percentage error was calculated using the formula $\frac{\eta^*_{asc}-\eta^*_{des}}{\eta^*_{asc}} \times 100\%$.

a. Data and percentage errors related to absolute dynamic viscosity $\eta^*(\omega)$ values measured using ascending frequency sweeps. From the dynamic light scattering we measured an average Mw of 306,000 g/mol.

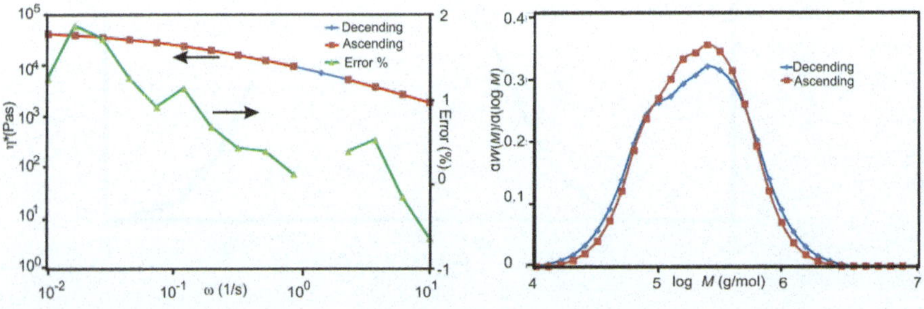

Fig. 14.8 Measured ascending or descending frequency sweeps for a PS sample and the corresponding computed MWDs. The differences between the results were greater at low frequencies, and data obtained when using an ascending frequency sweep yielded a more realistic MWD

b. Computed MWD results for PS exhibited less differences than for PE, but they were still visible. The MWD computed from data obtained using ascending frequency sweep was more realistic.

LCB may need much longer relaxation times for a new steady-state situation than do backbone chains. The relaxation state is dependent on the orientation of polymer chains and particularly on the amount of entanglement, with greater entanglement resulting in a longer time to obtain a new steady-state situation at a constant flow.

After each new step is reached during a frequency sweep, the stress changes are accepted too early when using the default preset standard tolerance settings of a rheometer, since the relaxation effect is still affecting the LCB. That is, the instrument can accept a new steady-state situation too quickly when using the default parameters to measure data points. For sheared and oriented chains, the relaxation takes longer during a descending frequency step due to the retardation time increasing the stress level. In contrast, during the steps in an ascending frequency sweep, the well-relaxed sample makes it easier to collect and detect information about LCB.

14.12 LCB and LCBD

The RED as the fingerprint of the polymer structure and properties represents an excellent rheological method for detecting LCB. Chrome-catalysed high-density PE, as shown in Fig. 14.9, is known to contain LCB, but cannot be detected by GPC/SEC, HPLC, or similar procedures. On the other hand, viscoelastic properties are very sensitive to LCB [75, 83, 84].

The results of simulating GPC from viscosity data η^*_{OBS} are indicated in Fig. 12.9 by the dotted lines for measured $\eta^*(\omega)$, $\eta^*_{OBS}(\omega)$ and w_{GPC}, for which the data at low frequencies are limited to the range $0.04/s < \omega < 40/s$. The RED is shown without smoothing. The red lines show both modelled Unified viscosity $\eta^*_{CHA}(\omega)$ and $w_{CHA}(\omega)$ computed from all of the $\eta^*_{OBS}(\omega)$ data measured, corresponding to a tenfold-wider frequency range of $0.004/s < \omega < 40/s$.

The influence of LCB on the RED is clearly evident. Note that the polymer sample and viscosity measurements were the same, with only the simulations being different. Differences for dynamic viscosity were minimal due to despite the use of a narrower data range for the GPC simulation, the influence of LCB was already physically included in the measured viscosity data and visible differences were only found outside the narrower measurement range $0.04/s < \omega < 40/s$ in the results shown in Fig. 14.9.

Figure 14.10 shows the converted LCBD and MWD obtained by melt calibration from the REDs shown in Fig. 14.9. We obtained the converted $w(M)$ from the second peak in the RED at 1,000,000 g/mol. This is surely affected by LCB and not HMW fractions, and thus we want to separate the LCBD from the complete $w(M)$. For MWD_{GPC} the average Mw was 165,242 g/mol and MwR was 14.6, which are close to the GPC/SEC results of $Mw = 170{,}000$ g/mol and $MwR = 14.6$. For the complete

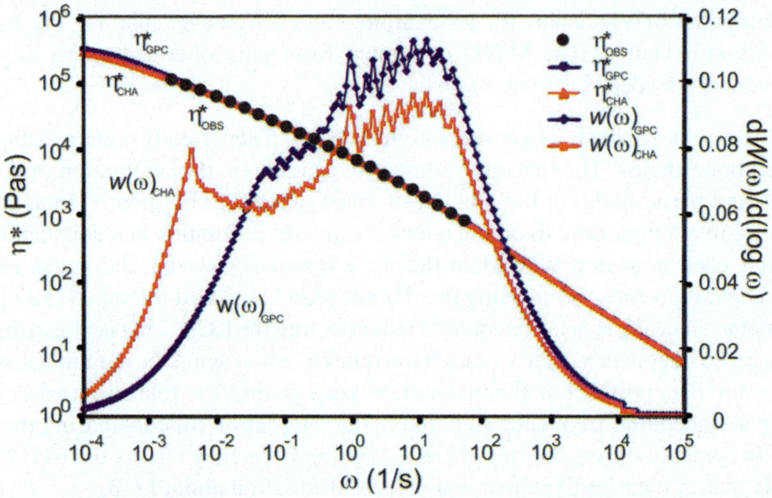

Fig. 14.9 Chromatograms or REDs and wide modelled viscosity curves for HDPE generated directly from viscosity measurements and by RheoAnalyzer computations. w_{GPC} (blue lines) corresponds to the results when using a narrower viscosity data range in simulating GPC/SEC measurements, as explained in the main text, while $w(\omega)$ (red lines) was obtained using a wider data range, which shows the second peak relating to LCB with same absolute viscosity value at characteristic frequency point $\omega_c = 1/s$ for demonstration purposes

$w(M)$ determined by melt calibration from the RED, the values were $Mw = 320{,}893$ g/mol and $MwR = 25.3$. The best value for separating $w_{LCB}(M)$ according to Eq. (14.4) was $k = 0.855$, which produced summation areas of 86% and 14% of the total amount for $\int w_{GPC}(M)$ and $\int w_{LCB}(M)$, respectively. According to Eq. (14.5) we obtained LCB frequencies of LCB/10^6 C = 0.81 and LCB/1000 C = 0.159.

The same values of Mf (polymer structure) and Hf (conversion factor) were used for the LCBD. Elongation viscosity experiments for strain hardening in Sect. 11.13 indicated that Mf and Hf must have totally different values in a start-up elongation experiment.

14.13 Conclusions

A new procedure for characterizing polymer structures using the MWD and LCB is presented herein based on a recently developed model. This computational rheological method has been introduced based on the terminology of liquid chromatography used with GPC/SEC and a new linear viscoelastic principle based on control theory. The method can describe relationships between the relaxation modulus, dynamic and shear viscosities, transient flow effects, temperature dependencies related to the RED and further by melt calibration of the MWD and LCB. The chromatogram or elugram measured by GPC/SEC relates only to the MWD, and the computed RED

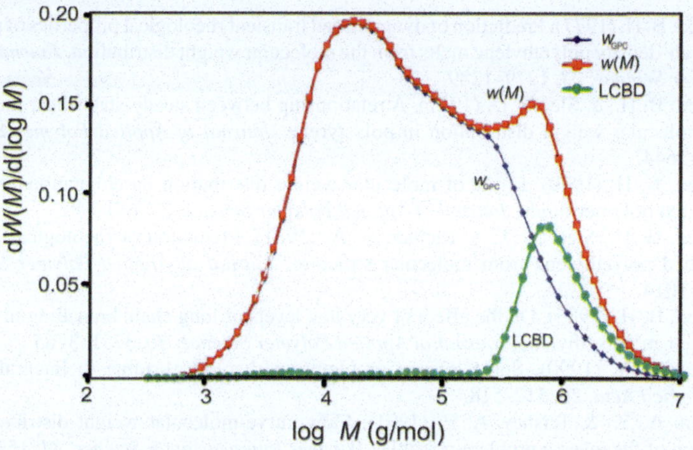

Fig. 14.10 MWDs and LCBD obtained from REDs shown in Fig. 12.9 by melt calibration for HDPE. Simulations of GPC/SEC measurements are indicated by w_{GPC} (blue line). The complete RED was converted into MWD $w(M)$ (red line), which shows the second peak relating to LCB. The tentative LCBD or $w_{LCB}(M)$ was obtained from the difference $w(M) - w_{GPC}(M)$. The real absolute values of the LCBD are still unclear

is simultaneously the accurate linear fingerprint of the structure and the rheological and viscoelastic properties. This RED fingerprint is also usable with mixtures comprising different chemical polymer structures and LCB. Analysis problems have been discussed specifically related to PE when obtaining the accurate MWD and the effects of rheometers performing frequency-sweep tests in an ascending or descending direction. The new procedure is presented in terms related to liquid chromatography and the few underlying mathematical formulas, and it has revealed the smoothing phenomena that occurs when applying GPC/SEC to PE. Example simulation results have been presented, with the results demonstrating that LCB and even the LCBD can be detected along with the MWD. In conclusion, computational rheological method represents a versatile novel method and tool for describing the flow properties and analyses of polymers.

Although many other polymers have been modelled, we have focused on polyolefins here to keep the presentation compact. For some special compounds, negative values for P' and P'' have been observed.

References

1. Bersted, B. H. (1975). An empirical model relating the molecular weight distribution of high-density polyethylene to the shear dependence of the steady shear melt viscosity. *Journal of Applied Polymer Science, 19*, 2167–2177.
2. Bersted, B. H. (1976). A model relating elastic properties of high-density polyethylene melts to the molecular weight distribution. *Journal of Applied Polymer Science, 20*, 2705–2714.

3. Bersted, B. H. (1977). Prediction of dynamic and transient rheological properties of polystyrene and high-density polyethylene melts from the molecular weight distribution. *Journal of Applied Polymer Science, 23*, 1279–1289.
4. Bersted, B. H., & Slee, J. D. (1977). A relationship between steady-state shear melt viscosity and molecular weight distribution in polystyrene. *Journal of Applied Polymer Science, 21*, 2631–2644.
5. Bersted, B. H. (1979). Effect of molecular weight distribution on elongational viscosity of undiluted polymer fluids. *Journal of Applied Polymer Science, 24*, 671–682.
6. Bersted, B. H., Slee, J. D., & Richter, C. A. (1981). Prediction of rheological behavior of branched polyethylene from molecular structure. *Journal of Applied Polymer Science, 26*, 1001–1014.
7. Bersted, B. H. (1985). On the effect of very low levels of long chain branching in rheological behavior in polyethylene. *Journal of Applied Polymer Science, 30*, 3751–3765.
8. Malkin, A. Y. (1990). Some inverse problems in rheology leading to integral equations. *Rheologica Acta, 29*, 512–518.
9. Malkin, A. Y., & Teishev, A. E. (1991). Flow curve-molecular weight distribution: Is the solution of the inverse problem possible? *Polymer Engineering & Science, 31*, 1590–1596.
10. Tuminello, W. H., & Cudré-Mauroux, N. (1991). Determining molecular weight distributions from viscosity versus shear rate flow curves. *Polymer Engineering & Science, 31*, 1496–1507.
11. Shaw, M. T., & Tuminello, W. H. (1994). A closer look at the MWD-viscosity transform. *Polymer Engineering & Science, 34*, 159–165.
12. Wood-Adams, P. M., & Dealy, J. M. (1996). Use of rheological measurements to estimate molecular weight distribution of linear polyethylene. *Journal of Rheology, 40*, 761–778.
13. Liu, Y., Shaw, M. T., & Tuminello, W. H. (1998). Obtaining MWD information from the viscosity data of linear polymer melts. *Journal of Rheology, 42*, 453–476.
14. Liu, Y., Shaw, M. T., & Tuminello, W. H. (1998). Optimized data collection for determination of the MWD from the viscosity data of polymer melts. *Polymer Engineering & Science, 38*, 169–176.
15. Liu, Y., & Shaw, M. T. (1998). Investigation of the nonlinear mixing rule for its adequacy in viscosity to molecular weight distribution transforms. *The Journal of Rheology*, *42*, 267–279.
16. Léonardi, F., Allal, A., & Marin, G. (1998). Determination of the molecular weight distribution of linear polymers by inversion of a blending law on complex viscosities. *Rheologica Acta, 37*, 199–213.
17. Tsenoglou, C. (1987). Viscoelasticity of binary homopolymer blends. *ACS Polym Prepr, 28*, 185–186.
18. Des Cloizeaux, J. (1988). Double reptation vs simple reptation in polymer melts. *Europhysics Letters, 5*, 437–442; *6*, 475.
19. Wasserman, S. H., & Graessley, W. W. (1992). Effects of polydispersity on linear viscoelasticity in entangled polymer melts. *Journal of Rheology, 36*, 543–572.
20. Cassagnau, P., Montfort, J. P., Marin, G., & Monge, P. (1993). Rheology of polydisperse polymers; relationship between intermolecular interactions and molecular weight distribution. *Rheologica Acta, 32*, 156–167.
21. Mavridis, H., & Shroff, R. J. (1993). Appraisal of a molecular weight distribution-to-rheology conversion scheme for linear polyethylenes. *Journal of Applied Polymer Science, 49*, 299–318.
22. Mead, D. W. (1994). Determination of molecular weight distributions of linear flexible polymers from linear viscoelastic material functions. *Journal of Rheology, 38*, 1797–1827.
23. Wasserman, S. H. (1995). Calculating the molecular weight distribution from linear viscoelastic response of polymer melts. *Journal of Rheology, 39*, 601–625.
24. Wasserman, S. H., & Graessley, W. W. (1996). Prediction of linear viscoelastic response for entangled polyolefin melts from molecular weight distribution. *Polymer Engineering & Science, 36*, 852–861.
25. Nobile, M. R., Cocchini, F., & Lawler, J. V. (1996). On the stability of molecular weight distributions as computed from the flow curves of polymer melts. *Journal of Rheology, 40*, 363–382.

26. Maier, D., Eckstein, A., Friedrich, C., & Honerkamp, J. (1998). Evaluation of models combining rheological data with the molecular weight distribution. *Journal of Rheology, 42*, 1153–1173.
27. Thimm, W., Friedrich, C., Marth, M., & Honerkamp, J. (1999). An analytical relation between relaxation time spectrum and molecular weight distribution. *Journal of Rheology, 43*, 1663–1672.
28. Thimm, W., Friedrich, C., Marth, M., & Honerkamp, J. (2000). On the Rouse spectrum and the determination of the molecular weight distribution from rheological data. *Journal of Rheology, 44*, 429–438.
29. Léonardi, F., Allal, A., & Marin, G. (2002). Molecular weight distribution from viscoelastic data: The importance of tube renewal and Rouse modes. *The Journal of Rheology, 46*, 209–224.
30. Vega, J. F., Rastogi, S., Peters, G. W. M., & Meijer, H. E. H. (2004). Rheology and reptation of linear polymers. Ultrahigh molecular weight chain dynamics in the melt. *The Journal of Rheology, 48*, 663–678.
31. Guzmán, J. D., Schieber, J. D., & Polland, R. (2005). A regularization-free method for the calculation of molecular weight distributions from dynamic moduli data. *Rheologica Acta, 44*, 342–351.
32. Nobile, M. R., & Cocchini, F. (2008). A generalized relation between MWD and relaxation time spectrum. *Rheologica Acta, 47*, 509–519.
33. Pattamaprom, C., Larson, R. G., & Sirivat, A. (2008). Determining polymer molecular weight distributions from rheological properties using the dual-constraint model. *Rheologica Acta, 47*, 689–700.
34. Friedrich, C., Loy, R. J., & Anderssen, R. S. (2009). Relaxation time spectrum molecular weight distribution relationships. *Rheologica Acta, 48*, 151–162.
35. Macskási, L. (2009). A critical analysis of the models connecting molecular mass distribution and shear viscosity functions. *Express Polymer Letters, 3*, 385–399.
36. Léonardi, F., Majesté, J.-C., Allal, A., & Marin, G. (2000). Rheological models based on the double reptation mixing rule: The effects of a polydisperse environment. *Journal of Rheology, 44*, 675–692.
37. van Ruymbeke, E., Keunings, R., & Bailly, C. (2002). Determination of the molecular weight distribution of entangled linear polymers from linear viscoelasticity data. *Journal of Non-Newtonian Fluid Mechanics, 105*, 153–175.
38. van Ruymbeke, E., Keunings, R., Stéphenne, V., Hagenaars, A., & Bailly, C. (2002). Evaluation of reptation models for predicting the linear viscoelastic properties of entangled linear polymers. *Macromolecules, 35*, 2689–2699.
39. van Ruymbeke, E., Stéphenne, V., Daoust, D., Godard, P., Keunings, R., & Bailly, C. (2005). A sensitive method to detect very low levels of long chain branching from the molar mass distribution and linear viscoelastic response. *Journal of Rheology, 49*, 1503–1520.
40. Jam, J., Nekoomanesh, M., & Ahmadi, M. (2012). From molecular weight distribution to linear viscoelastic properties and back again: Application to some commercial high-density polyethylenes. *Iranian Polymer Journal, 21*, 403–413.
41. Farias, T. M., Cardozo, N. S. M., & Secchi, A. R. (2013). The use of Gauss-Hermite quadrature in the determination of the molecular weight distribution of linear polymers by rheometry. *Brazilian Journal of Chemical Engineering, 30*, 909–921.
42. Cocchini, F., & Nobile, M. R. (2003). Constrained inversion of rheological data to molecular weight distribution for polymer melts. *Rheologica Acta, 42*, 232–242.
43. Trathnigg, B. (2000). Size-exclusion chromatography of polymers. In R. A. Meyers (Ed.), *Encyclopedia of analytical chemistry* (pp. 8008–8034). John Wiley & Sons Ldt.
44. Mori, S., et al. (1996). Report on cooperative determination of molecular-weight averages of polymers by size-exclusion chromatography. 2. A Report on the Second Round-robin Test (No. 1). *Bun-seki Kagaku, 45*(1), 95–101.
45. Bruessau, R. J. (1996). Macromolecular symposia. *Experiences with Interlaboratory GPC Experiments, 110*, 15–32.
46. Lederer, K., & Aust, N. J. (1996). Molecular characterization of polyolefins. *Journal of Macromolecular Science, Pure and Applied Chemistry, 33*, 927–940.

47. D'Agnillo, L., Soares, J. B. P., & Penlidis, A. (2002). Round-Robin experiment in high-temperature gel permeation chromatography. *Journal of Polymer Science Part B: Polymer Physics, 40*, 905–921.
48. Ritter, A., Schmid, M., & Affolter, S. (2010). Determination of molecular weights by size exclusion chromatography (SEC)—results of round robin tests. *Polymer Testing, 29*, 945–952.
49. Luruli, N. (2010). Repeatability and Reproducibility of Sample Preparation and Analysis in High-Temperature SEC, IUPAC.
50. Laun, M., Auhl, D., Brummer, R., Dijkstra, D. J., Gabriel, C., Mangnus, M. A., Rüllmann, M., Zoetelief, W., & Handge, U. A. (2014). Guidelines for checking performance and verifying accuracy of rotational rheometers: Viscosity measurements in steady and oscillatory shear. *Pure and Applied Chemistry, 86*, 1945–1968.
51. Jiang, W., Sun, Y., Xu, Y., Peng, C., Gong, X., & Zhang, Z. (2010). Shear-thickening behavior of polymethylmethacrylate particles suspensions in glycerine–water mixtures. *Rheologica Acta, 49*, 1157–1163.
52. Lee, Y. S., & Wagner, N. J. (2003). Dynamic properties of shear thickening colloidal suspension. *Rheologica Acta, 42*, 199–208.
53. Dealy, J. M., & Larson, R. G. (2006). *Structure and rheology of molten polymers, from structure to flow behavior and back again.* Carl Hanser Verlag.
54. Graessley, W. W. (2003). *Polymeric liquids and networks: Structure and properties.* Garland Science.
55. Graessley, W. W. (2008). *Polymeric liquids and networks: Dynamics and rheology.* Garland Science.
56. Striegel, A. M., Yau, W. W., Kirkland, J. J., & Bly, D. D. (2009). *Modern size-exclusion liquid chromatography.* Wiley.
57. Dorsey, J. G., Cooper, W. T., Siles, B. A., Foley, J. P., & Barth, H. G. (1998). Liquid chromatography: Theory and methodology. *Analytical Chemistry, 70*, 591R.
58. Schroder, E., Muller, G., & Arndt, K. F. (1988). *Polymer characterization.* Hanser Publishers.
59. Dong, M. W. (2006). *Modern HPLC for practicing scientists.* Wiley.
60. Ettre, L. S. (1993). Nomenclature for chromatography (IUPAC Recommendations 1993). *Pure and Applied Chemistry, 65*(4), 819–872.
61. Fleury, G., Schlatter, G., & Muller, R. (2004). Non linear rheology for long chain branching characterization, comparison of two methodologies: Fourier transform rheology and relaxation. *Rheologica Acta, 44*, 174–187.
62. Malmberg, A., Gabriel, C., Steffl, T., Münstedt, H., & Löfgren, B. (2002). Long-chain branching in metallocene-catalyzed polyethylenes investigated by low oscillatory shear and uniaxial extensional rheometry. *Macromolecules, 35*, 1038–1048.
63. Costeux, S. (2003). Modeling of randomly branched polymers produced by combination of several single-site catalysts: Toward optimization of melt properties. *Macromolecules, 36*, 4168–4187.
64. Chen, X., Costeux, C., & Larson, R. G. (2010). Characterization and prediction of long-chain branching in commercial polyethylenes by a combination of rheology and modeling methods. *Journal of Rheology, 54*, 1185–1206.
65. Wood-Adams, P. M., Dealy, J. M., DeGroot, A. W., & Redwine, O. D. (2000). Effect of molecular structure on the linear viscoelastic behavior of polyethylene. *Macromolecules, 33*, 7489–7499.
66. Sugimoto, M., Suzuki, Y., Hyun, K., Ahn, K. H., Ushioda, T., Nishioka, A., Taniguchi, T., & Koyama, K. (2006). Melt rheology of long-chain-branched polypropylenes. *Rheologica Acta, 46*, 33–44.
67. Vega, J. F., Expósito, M. T., Martínez-Salazar, J., Lobón-Poo, M., Barcina, J. O., Martínez, A. G., & López, M. (2011). Molecular architecture and linear viscoelasticity of homogeneous ethylene/styrene copolymers. *Rheologica Acta, 50*, 207–220.
68. Zimm, B. H., & Stockmayer, W. H. (1949). The dimensions of chain molecules containing branches and rings. *The Journal of Chemical Physics, 17*, 1301–1314.

69. Gabriel, C., & Münstedt, H. (2002). Influence of long-chain branches in polyethylenes on linear viscoelastic flow properties in shear. *Rheologica Acta, 41*, 232–244.
70. Wood-Adams, P. M., & Dealy, J. M. (2000). Using rheological data to determine the branching level in metallocene polyethylenes. *Macromolecules, 33*, 7481–7488.
71. van Gurp, M., & Palmen, J. (1998). Time temperature superposition of polymeric blends. *Rheology Bulletin, 67*, 5–8.
72. Trinkle, S., Walter, P., & Friedrich, C. (2002). Van Gurp–Palmen Plot II—classification of long chain branched polymers by their topology. *Rheologica Acta, 41*, 103–113.
73. García-Franco, C. A., Srinivas, S., Lohse, D. J., & Brant, P. (2001). Similarities between gelation and long chain branching viscoelastic behavior. *Macromolecules, 34*, 3115–3117.
74. Keßner, U., Kaschta, J., & Münstedt, H. (2009). Determination of method-invariant activation energies of long-chain branched low-density polyethylenes. *Journal of Rheology, 53*, 1001–1016.
75. Kessner, U., Kaschta, J., Stadler, F. J., Le Duff, C. S., Drooghaag, X., & Münstedt, H. (2010). Thermorheological behavior of various short- and long-chain branched polyethylenes and their correlations with the molecular structure. *Macromolecules, 43*, 7341–7350.
76. Malmberg, A., Liimatta, J., Lehtinen, A., & Löfgren, B. (1999). Characteristics of long chain branching in ethene polymerization with single site catalysts. *Macromolecules, 32*, 6687–6696.
77. Malmberg, A., Kokko, E., Lehmus, P., Löfgren, B., & Seppälä, J. (1998). Long-chain branched polyethene polymerized by metallocene catalysts Et[Ind]2ZrCl2/MAO and Et[IndH4]2ZrCl2/MAO. *Macromolecules, 31*, 8448–8454.
78. Gabriel, C., Kokko, E., Löfgren, B., Seppälä, J., & Münstedt, H. (2002). Analytical and rheological characterization of long-chain branched metallocene-catalyzed ethylene homopolymers. *Polymer, 43*, 6383–6390.
79. Soares, J. B. P., & Hamielec, A. E. (1996). Bivariate chain length and long chain branching distribution for copolymerization of olefins and polyolefin chains containing terminal double bonds. *Macromolecular Theory and Simulations, 5*, 547–572.
80. Meimaroglou, D., Krallis, A., Saliakas, V., & Kiparissides, C. (2007). Prediction of the bivariate molecular weight-long chain branching distribution in highly branched polymerization systems using Monte Carlo and sectional grid methods. *Macromolecules, 40*, 2224–2234.
81. Meimaroglou, D., & Kiparissides, C. (2010). A novel stochastic approach for the prediction of the exact topological characteristics and rheological properties of highly-branched polymer chains. *Macromolecules, 43*, 5820–5832.
82. Costeux, S., Wood-Adams, P., & Beigzadeh, D. (2002). Molecular structure of metallocene-catalyzed polyethylene: Rheologically relevant representation of branching architecture in single catalyst and blended systems. *Macromolecules, 35*, 2514–2528.
83. Lohse, D. J., Milner, S. T., Fetters, L. J., Xenidou, M., Hadjichristidis, N., Mendelson, R. A., Garcia-Franco, C. A., & Lyon, M. K. (2002). Well-defined, model long chain branched polyethylene. 2. Melt rheological behavior. *Macromolecules, 35*, 3066–3075.
84. Vega, J. F., Santamaría, A., Muñoz-Escalona, A., & Lafuente, P. (1998). Small-amplitude oscillatory shear flow measurements as a tool to detect very low amounts of long chain branching in polyethylenes. *Macromolecules, 31*, 3639–3647.

Chapter 15
Hemorheological Method for Analyses

This chapter is based on a non-published peer-reviewed manuscript. A new hemorheological procedure for analysing the structure and viscoelasticity of blood components is presented. The analysis method presented herein can use viscosity data obtained from different types of blood flow, such as plasma and serum, or blood as it is. This chapter is focused on the viscosity data obtained from a shear, oscillating rheometer or respective device to determine structural information also related to electropherogram and results of capillary electrophoresis.

15.1 Introduction

Recent studies [1–4] have found that blood behaves similarly to polymers, in which viscoelasticity is well known by traditional Maxwell-type models first composited by Gross [5] in the so called "relaxation time principle" and later thoroughly by others [6, 7]. Blood proteins tend to return to their original shape after their chainlike molecules are stretched by the flow related to standard orientation of polymer chains. The viscoelastic behaviours of polymers have been studied for a long time as they are easier to detect, measure, and study, and numerous rheological models have been developed over the years. The viscoelastic behaviours of polymers have been studied for a long time as they are easier to detect, measure, and study, and numerous rheological models have been developed over the years. The methods and software of rheology are introduced briefly.

A brief overview of the background and the use of control theory is reported earlier in Chap. 2. Detailed explanations about the background principles and formulas have previously been presented for different flows by Unified Model; steady shear viscosity in Chap. 4, the dynamic viscosity in Chap. 5, the relaxation modulus in Chap. 6, transition viscosity in Chap. 10 and elongation viscosity in Chap. 11. We repeat here only the information needed for practical blood analyses based on shear

T. Borg, *Control Theory in Rheology*, https://doi.org/10.1007/978-3-031-88248-7_15

viscosity data. As the background of the constitutive and mathematical foundation provided by control theory for polymer rheology and analyses is a broad subject discussed in earlier chapters, we limit our presentation here to a new standpoint of the analysis procedure converted to hemorheological uses.

We introduce and use some variables and names from polymer theory to help the reader, if they want to study the respective procedures used for decades in polymers. For polymer chains, Molar Mass Distribution (MMD) is used, and the mathematical form is the normalized function $w(\log M)$. Another term used is Molecular Weight Distribution (MWD). The formula used for shear viscosity $\eta\,(\dot{\gamma})$ as a function of shear rate is normally used. Although some measurements are done for complex viscosity η^* as a function of frequency ω using an oscillation rheometer and many polymer analyses, here we present all results by shear viscosity and rate. Principle, formulas and software are easily converted to another scales. For calibration purposes, we first convert the electropherogram to a viscosity flow curve and back, which is possible using the linear principle of analysis. The word "linear" have a deeper meaning in mathematics and physics than a simple straight line.

We start explaining the method by converting principle from electropherogram to viscosity flow curve. Next, we demonstrate the conversion from viscosity to composite distribution, and then back to the electropherogram. Finally, computations and suggestions for measurement methods are discussed from the standpoint of the principle. This preliminary study of the new method is kept compact as several test measurements are still needed to further develop the analysis procedures.

15.2 From Blood Structure to Viscosity

First, a step-by-step procedure from electropherogram to model shear viscosity $\eta(\dot{\gamma})$ and other viscoelastic properties is presented. This procedure is needed to set up the analysis process and is done continuously internally during computations. Figure 15.1 shows the development route from electropherogram to viscosity flow curve, and we use the respective alphabet in the text sections.

15.2.1 An Electropherogram and Definition for MMD_a

Figure 15.1a shows a typical normal pattern for a capillary electropherogram of human serum proteins, but we can also use plasma or complete blood for this procedure. Because serum is a blend of different components such as proteins, ions, nutrients, wastes, water, and dissolved gases, we cannot use the same MWD definition for blood as we do for polymers, which consist of similar chemical-type molecule chains of different lengths.

Blood consists of different types of chemical and molecular structures. The Unified method requires a single distribution function, even with several pattern tops.

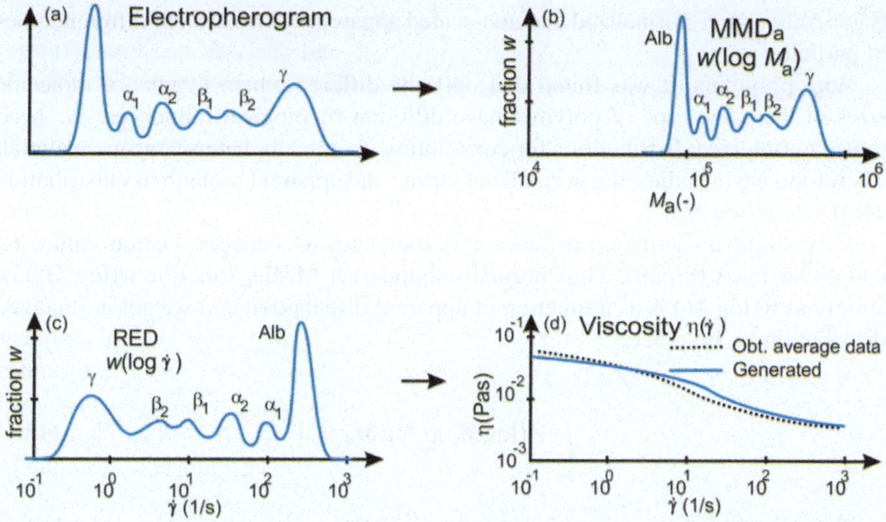

Fig. 15.1 Schematic diagrams explaining how electropherogram is converted to viscosity curve

We introduce the virtual or apparent molar mass distribution (MMD_a). Figure 15.1 a presents a typical normal pattern for serum protein components in blood serum: Albumin (Alb), $Alpha_1$ (α_1), $Alpha_2$ (α_2), $Beta_1$ (β_1), $Beta_2$ (β_2) and Gamma (γ) fractions without delving deeper into their structures. In this study we do not discuss absolute total values of fractions in grams per decilitre (g/dl) or respective scales.

For this model, the x-axis on the electropherogram must be scaled to some variable that arranges the composition particles and fractions properly as a function of molar mass, molecular chain length, particle dimension, or some other respective feature. This property must be linear on the x-axis, as is the case with melt calibration of polymers or as needed in the Universal calibration curve used with gel permeation chromatography (GPC), also referred to as size-exclusion chromatography (SEC) methods. The literature provides some average molecular weights *M*w (g/mol) or molar mass in Daltons, abbreviated as Da (–). For illustration on charts, we also use a logarithmic form for most scales related to the equations. Some average data is found in the literature in Daltons: Alb = 66,500 – 69,000 (log 4.83–log 4.84), $Alpha_1$ (α_1) = 72,000 (log 4.86), $Alpha_2$ (α_2) = 85,000 (log 4.93), Gamma (γ) = 340,000 (log 5.53).

15.2.2 Normalization of an Electropherogram

In Fig. 15.1b, we perform some conversions. With the Unified principle and formulas, it is advantageous to handle distributions in normalized form, and the cumulative distribution must be normalized for integral equations. A typical electropherogram is generally a non-normalized and non-scaled apparent distribution of different types of particles.

With polymers, it was found that only the different chain lengths or molecule sizes of the same type of polymer have different rheological influences. We need to use normalized distributions for constituting chemically heterogeneous material. Therefore, we introduce the normalized virtual and apparent molar mass distribution MMD_a or $w(\log M_a)$.

Very slight insignificant difference is found for M between Dalton values (-) and molar mass (g/mol). Thus normalised apparent MMD_a function $w(\log M_a)$ or simply as $w(\log M)$ is as a function of apparent distribution and we get normalized distribution by Eq. 15.1:

$$\int_{-\infty}^{\infty} w(\log M_a) d \log M_a = 1 \tag{15.1}$$

Complete blood consists of red blood cells (RBCs), or erythrocytes, which have a dramatic influence on the absolute viscosity level, as discussed later. In this presentation, we use serum, and apparent distributions for plasma or complete blood with RBCs are discussed later.

Another advantage of this principle is that in plasma or serum, red blood cells (RBCs) are separated beforehand, making it more practical for viscosity measurements and structure analyses with this method. A significant feature is that this principle uses relative viscosity values, not absolute ones. Whether RBCs are present or not has no influence on the structure analyses, as explained later.

15.2.3 Apparent Distribution Function to a Rheological RED

In Fig. 15.1c, we convert a material structure function to the respective viscoelastic property function in a manner analogous to methods used with polymers. By involving the chromatogram and our melt calibration, we can convert a material structure function to the respective viscoelastic property function. Melt calibration, presented in Chap. 7 for polymers, is the relation between rheological properties and molecular weight. It allows the conversion to any linear rate or time scales by adjusting the structural value *Mf* and scale units using the conversion factor *Hf*.

An important practical point is that the procedure requires only one reference sample with a known average molecular weight, which is set by the structural value *Mf*. Similarly, the polydispersity value $Đ_\text{M}$ (*MwR*) is calibrated by the *Hf* value. Thus, the melt calibration curve is a straight line on a logarithmic scale, similar to how the GPC/SEC calibration curve was generated by the Mark-Houwink equation and further Benoit's Universal Calibration curves. In analyses, our melt calibration with only two constants, *Mf* and *Hf*, performs well.

A rheometer adds mechanical energy to a sample to generate shear viscosity in the form of stress as a function of rate as a response, from which the rheologically effective distribution (RED) and further electropherogram are computed. The RED is the fingerprint of the molecular and dynamic structure and of viscoelastic properties.

For blood analyses, we can convert the function of the apparent molar mass distribution $w(\log M_a)$ into a rheologically effective distribution (RED) of the shear rate $w(\log \dot{\gamma})$ using a conversion factor between the scale *Hf* and the polymer structural value *Mf*. We have developed a simple relation between property and structure functions, $M(\dot{\gamma})$, as a function of shear rate, where the value of *Mf* is *M* at $\dot{\gamma}_1 = 1/$ s:

$$M = Mf\left(\frac{\dot{\gamma}_1}{\dot{\gamma}}\right)^{\frac{1}{Hf}} \tag{15.2}$$

For polymer measurements, the *Hf* conversion factor is usually close to 4 (with the exception of long and thin chain structure PE, whose values are $2 < Hf < 3$), and *Mf* takes values in Daltons between 10^2 and 10^6. For human blood we found with preliminary computations in range $Hf = 2.72$ and $Mf = 350{,}000$.

Equations. (15.2) provides the linear calibration line for a logarithmic scale, which has worked well so far for this relationship. It is analogous to liquid chromatography methods such as GPC, also referred to as SEC, and high-performance liquid chromatography (HPLC). If a different form of a curved function is found to be a better alternative, it can easily be substituted, as is done with GPC/SEC systems. Now, the RED $w(\log \dot{\gamma})$ function is insensitive to different forms of particle structures in the subsequent steps, so we do not need to use the term "apparent." The RED is the direct linear fingerprint of the molecular and dynamic structure and of viscoelastic properties such as the relaxation modulus or the dynamic and shear viscosities. We must again normalize *w* to ensure that Eq. 15.3 is correct, following the procedures presented in Chap. 3.

$$\int_{-\infty}^{\infty} w(\log \dot{\gamma}) d \log \psi = 1 \tag{15.3}$$

15.2.4 From RED to Viscosity Flow Curve

In Fig. 15.1d viscosity flow curve η is generated easily from $w(\log \dot{\gamma})$ RED distribution using Eq. 15.4, which will be discussed later. Before that, we need to discuss the measured viscosity data of blood from different studies, the devices used, and the effects of red blood cells (RBCs).

15.3 Red Blood Cells (RBCs)

The literature reveals variations in measured viscosity data, especially between measurements obtained using capillary or rotational rheometers, and also across different studies over the years. In a comprehensive review, Sousa et al. [8] compiled blood viscosity measurements from various authors under steady shear conditions.

The document presents data at different hematocrit (RBC) levels, ranging from 41.2 to 51.7%, measured at physiological temperature (37 °C) using diverse methodologies, primarily involving various types of capillary rheometers. The authors caution that due to the variety of conditions employed in different studies—such as hematocrit levels, measurement temperatures, shear rate ranges, and blood sample preparation methods—direct comparisons between the results are challenging. Alves et al. [9] conducted viscosity measurements using a cone-and-plate rheometer at different rotational shear rates, utilizing a controlled stress rheometer in rotating mode. They excluded values obtained at the highest and lowest shear rates, noting an abnormal increase in viscosity at higher shear rates.

Recent articles [10–12] have compiled data that contribute to our composite flow curve, representing an average result for human blood at a temperature of + 37 °C. This data is presented in classical, non-logarithmic form in Fig. 15.2. Throughout this document, it is utilized in simulations and referred to as the apparent measured viscosity, treated as true measured data.

Thus, red blood cells, or erythrocytes, have a dramatic influence on the absolute viscosity level. Similarities with polymers were found in that long chain branching (LCB) and now red blood cells (RBC) affect the viscosity at shear rates somewhere below 0.01/s (10^{-2}/s), significantly increasing the overall viscosity. Moreover, larger red blood cells (RBC) appear to exhibit rubber-like particle behaviour, similar to some fillers in polymer melts.

Thus, the local point influence of RBCs on the RED is entirely outside the measurement range, but the principle and software still detect the phenomenon. As a result, the amount of RBCs has a fundamental influence on the overall absolute blood viscosity values, both in principle and in practice. RBCs act as filler blocks, setting the zero viscosity level and creating local relative effects outside the measuring range.

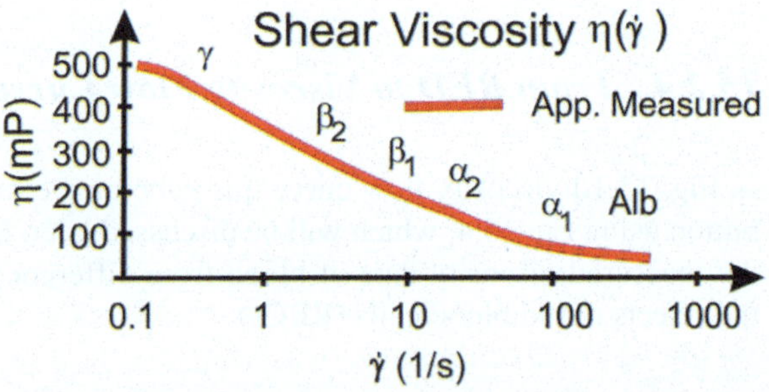

Fig. 15.2 The apparent measured blood viscosity is collected from literature and treated as true measured data. The same used viscosity data is shown in Fig. 15.1d, converted to the classically presented scales in hemorheology, with the same blood structure fractions added

Additional influences of the RBCs include the Fåhræus effect [13] and the Fåhræus-Lindqvist effect [14], which occur at higher shear rates, typically in the range of 500/s to 1000/s, depending on the measuring device and rheometer.

15.4 Unified Formula for a Viscosity Flow Curve

Finally we can explain the procedure to manually get the viscosity fit curve $\eta(\dot{\gamma})$ using Eq. 15.4. We initially found that the RED $w(\log \dot{\gamma})$ distribution consistently exhibits a strongly bimodal structure for viscoelasticity, which led us to consider partitioning it into two REDs. Strictly from the standpoint of the constitutive equations, the distribution components should be named RED', obtained by flow measurements, and RED", used to achieve the model fit.

The normalized RED is partitioned using elasticity value P' or scaled $P'w'$ (log $\dot{\gamma}$) to obtain RED' or $P''w''(\log \dot{\gamma}/R)$, respectively, where R is the distance for the same point for the effective w' (log $\dot{\gamma}$) and w"(log $\dot{\gamma}$). For these narrow distributions of blood it can be as initial constant $R = 10^6$. Since effective changes in the viscosity component typically occur at $\dot{\gamma} < 0.001$/s (which is outside the measurement range) and the form of the RED" only minimally affects the obtained relaxation, the use of $w''(\log \dot{\gamma}/R)$ for the same formula with a similar form is the best available choice in the absence of more information.

$$\log \frac{\eta(\dot{\gamma})}{\eta_0} = -\log \frac{\dot{\gamma}}{\dot{\gamma}_c} \int_{-\infty}^{\log \dot{\gamma}} \left(P'w'(\log \psi) + P''w''\left(\log \frac{\psi}{R}\right)\right) d \log \psi \qquad (15.4)$$

Adopting relative values introduces another beneficial feature. For the electropherogram and the construction of MMD_a and RED data of serum, we utilize the viscosity of complete blood. Importantly, the hematocrit level fundamentally influences the absolute viscosity values.

Therefore, we employ a characteristic viscosity value, η_0, at a characteristic shear rate, $\dot{\gamma}_c = 1/s$, for blood, which establishes the absolute level of the complete viscosity flow curve. However, you will still need to visually determine the P' and P'' values to ensure your model aligns with the measurements. This approach also allows you to model viscosities for new REDs or other parameters that have not yet been observed. All these tasks are performed numerically by the software described in the following section.

Although the same modelled viscosity flow curve is depicted across all figures up to the more detailed Fig. 15.4, it is challenging to visually discern the minimal bends and twists in the viscosity curve caused by the effects of composite tops.

The computation in the opposite direction is explained, starting from the viscosity flow curve. Figure 15.3 shows the development route to structure analysis, electropherogram, RED, and other viscoelastic functions. The procedure explained here in Figs. 15.1 and 15.3 is repeated around a thousand times during computations for measured viscosity data, with iterations aiming to find the best fits for variables and measured data, as presented in Chap. 9.

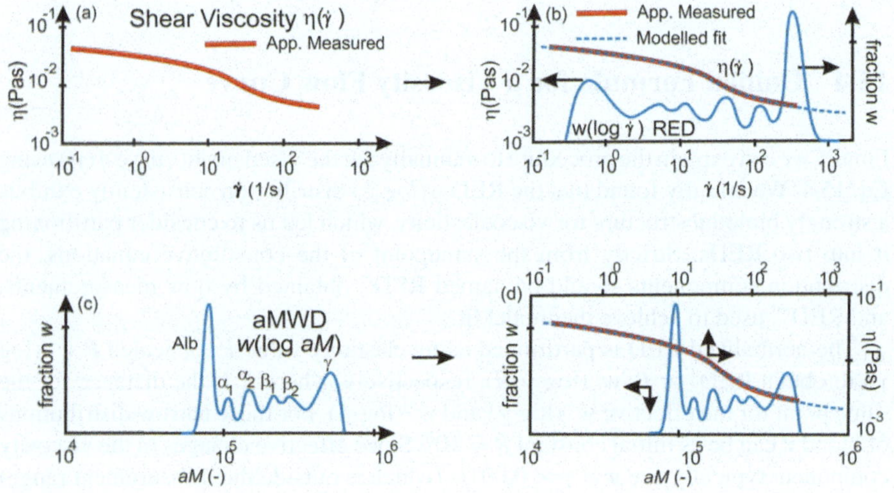

Fig. 15.3 Schematic diagrams explaining the conversion of the viscosity curve to the electropherogram, molar mass distribution, and software outputs

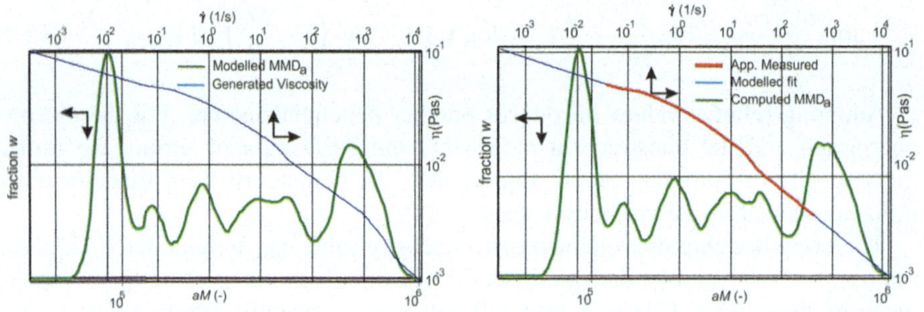

Fig. 15.4 On the left software print out as it is of electropherogram related mathematically linearly to the viscosity flow curve. On the right is molar mass distribution related mathematically linearly to the viscosity flow curve

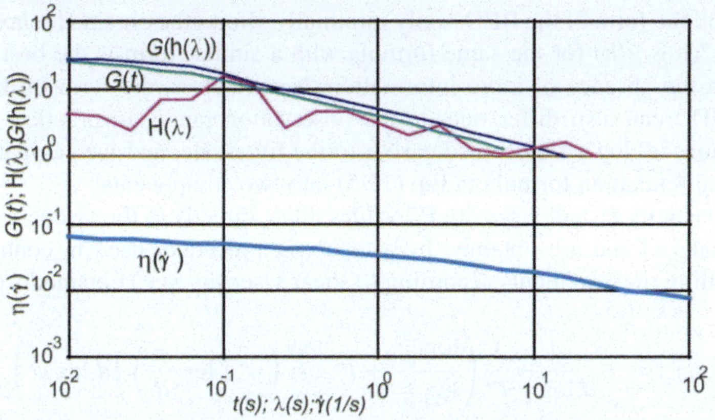

Fig. 15.5 Software print out for relaxation modulus $G(t)$ and spectrum $H(\lambda)$ and again computed relaxation modulus $G[H(\lambda)]$ from spectrum. Relaxation modulus is an important factor with polymers informing resistance and relaxation properties of matter is sudden transformation of state. Transition rate is at highest at 0.1s after impulse obtained from relaxation spectrum $H(\lambda)$, which suits well for the heart pulse rate

15.5 From Viscosity to MMD_a and Electropherogram

15.5.1 Starting from Viscosity Flow Curve

Normally, analyses begin at this point with the measured viscosity flow curve, but we did not yet have data with enough datapoints suited for this method. Therefore, we use a section of the viscosity curve from the earlier model shown in Fig. 15.3a for simulation purposes, treating it as the measured viscosity curve. In practice, the modelled viscosity flow curve is compared with the measured data, and numerical and artificial intelligence procedures employed by the computation software find the best fits. This procedure is rather ill-posed and sensitive to errors. Therefore, it is impossible to manually smooth any data afterward that is electronically recorded directly from the rotation head or other sensors, as discussed in Chap. 9 and elsewhere.

15.5.2 By Deriving We Get RED

In Fig. 15.3b the normalized RED is partitioned using *P'* or scaled *P'w'* (log $\dot{\gamma}$) to obtain RED' or *P"w"* (log $\dot{\gamma}/R$), respectively, where *R* is the distance for the same point for the effective *w'* (log $\dot{\gamma}$) and *w"*(log $\dot{\gamma}$). Sum value *P* is determined by a feedback procedure from partitioned *P'* and *P"* values, or as $P = P' + P''$, and in practice can accurately represent polymer flows. Since effective changes in the viscosity force typically occur at $\dot{\gamma} < 0.001$ s (which is outside the measurement range) and the form of the RED" only minimally affects the obtained relaxation, the use of $w''(\log \dot{\gamma}/R)$ for the same formula with a similar form is the best available choice in the absence of more information. Nevertheless, the form of the $w''(\log \dot{\gamma}/R)$ RED" can also differ outside the observation range. In practice, the same distribution, w' log $\dot{\gamma}$), is copied earlier to the timescale, and we can partition the Analytical relaxation formula in Eq. (15.5) into two components:

The elasticity P' and viscosity P'' values relate directly to the viscoelastic nature of the material and are obtained by a feedback procedure used in control theory during fitting measurements according to shear viscosity $\eta(\dot{\gamma})$ obtained from

$$w'(\log \dot{\gamma}) = -\frac{d}{d \log \dot{\gamma}} \frac{1}{P'} \left(\frac{\log \frac{\eta}{\eta_0}}{\log \frac{\dot{\gamma}}{\dot{\gamma}_c}} + P'' \int_{-\infty}^{\log \dot{\gamma}} \left(w''\left(\log \frac{\psi}{R} \right) \right) d \log \psi \right) \quad (15.5)$$

The formulas fundamentally contain a variable in the exponent, and thus in numerical computation practice, we need to consider logarithmic and exponential functions simultaneously during cyclic derivation, integration, and other computations. This power-integral-logarithm-derivation processing is a very unstable technique due to the differences between the measurements and the model values. Therefore, we adapted standard numerical methods found in textbooks and further developed them for this specific case.

This loop between Eqs. (15.4) and (15.5) is performed around a thousand times during a single computation to obtain the best feedback fits to the RED and viscosity function according to the measured data. In summary, we obtain linear relations that are accurate within the numerical rounding errors between the RED and shear viscosity η ($\dot{\gamma}$).

15.5.3 Apparent Distribution Function

In Fig. 15.3c Normalised apparent MMD_a $w(\log M_a)$ is converted by Eq. (15.2), but to another direction. We point again that apparent distribution consist very different type structures, but some molar mass fractions were known and usable in calibration of RED. Now gained and normalized apparent MMD_a $w(\log M_a)$ is an electropherogram, if the scale of x-axes is omitted.

15.5.4 Modelled Viscosity and Other Viscoelastic Properties

In Fig. 15.3d is shown that during computation is always modelled wide range viscosity fit curve. For polymers is developed procedures to model relaxation modulus $G(t)$ and relaxation spectrum $H(\lambda)$ presented in Chap. 6. For blood analyses they may not be important, but are created easily by software simultaneously.

The relaxation spectrum $H(\lambda)$ can be approximated by deriving relaxation modulus $G(t)$ according to the procedures presented by Tschoegl [6]. We need relaxation modulus $G(t)$ data in modelling blood flows and viscosity at different shear rates called transition viscosity.

Figure 15.4 presents, on the left, the same result as shown in Fig. 15.1d. This result was computed by the RheoDeveloper software, starting from an electropherogram and apparent MMD_a $w(\log M_a)$ to derive the viscosity curve $\eta(\dot{\gamma})$. Used elasticity value $P' = 0.37$ and viscosity value $P'' = 0.39$, $Mf = 350{,}000$ and $Hf = 2.72$. Average molar mass $Mw = 370{,}900$ and polydispersity index is rather narrow $Đ_M = 1.74$.

On the right is output of RheoAnalyzer software starting from modelled and recorded viscosity data of RheoDeveloper software. At high end on the right of MMD_a are light distortions, but computed viscosity flow curve has only small error with original or root mean square error RMSE% = 0.0026.

15.6 Relaxation Modulus *G*(*t*) and Spectrum *H*(λ)

For the computed by Unified Model relaxation modulus, $G(t)$ is used formula Eq. 13.6 of principle:

$$\log \frac{G(t)}{G_0} = -\log \frac{t}{t_c} \int_{-\infty}^{\log t} \left(P'w'(\log \tau) + P''w''\left(\log \frac{\tau}{R}\right)\right) d \log \tau \tag{15.6}$$

The characteristic constant t_c normally has a value of 1 s, resulting from the shifted impulse response during the approximation procedure. By standard simple numerical fitting procedures, such as bracketing, the zero modulus G_0 is determined. A constant characteristic time of $\tau_c = 10^{-6}$ s was used in the relaxation simulations. The spectrum's maximal relaxation time, λ max, was set to 400 s.

The flow of blood is cyclic according to heart pulses and different flow rates. According to Alves and co-authors [9] wrote that Kensey et. Al . [15] and for a typical iliac artery, the shear rate varies from approximately 500/s to zero during a cardiac cycle. Other studies [16–18] report that instantaneous shear rate, over a cardiac cycle, varies from zero to approximately 1000/s in several large arterial vessels. Relaxation modulus relates transition rate of viscosity at different shear rates at new flow rate. These results come simultaneously out during computation, but results are shown later on with real measured viscosity data analyses. The relaxation speed of blood is maybe more important than viscosity as a function of shear rate. Relaxation modulus G(t) and spectrum $H(\lambda)$ are shown in Fig. 15.5.

15.7 Suggested Measuring Routines

This document was written using apparent average viscosity measured by different types of rheometers. For polymers, viscoelastic data is typically measured by an oscillation rheometer, with complex viscosity data being used primarily due to instrumental reasons. For polymers, frequencies ω are more desirable at higher values for well-relaxed samples, with the oscillation rheometer at small strain providing five data points per rate decade. In this study, frequencies with the oscillation rheometer are used as the general shear rate $\dot{\gamma}$.

For modelling flows in a tube or blood vessel, absolute viscosity values are crucial; however, the principle requires relatively accurate data for analytical purposes. For blood components, a much narrower range of $0.01/\text{s} < \omega < 100/\text{s}$ is necessary, but more data points per rate decade are essential to detect component fractions as comprehensively as possible. For polymers, it is customary to use logarithmic steps of 0.2 in frequencies (ω), yielding five data points per decade. For thorough blood analysis, at least logarithmic steps of 0.05 in frequencies or shear rates are needed, providing twenty data points per decade. The most precise results are achieved with logarithmic steps of 0.02.

We found in this study that for plate-plate head is needed a conversion factor *Hf* in range value 2.0 and cone-plate respectively $2.7 < Hf < 3.0$. The preliminary values are as at same level as for long chain polyethylene melts.

As blood is a liquid, a cone-plate head may not be the optimal configuration for a rotating rheometer; instead, a concentric cylinder design may be more suitable. We are primarily interested in measuring some form of resistance force at different rotating speeds, which should be measured relatively accurately on a linear scale. [19].

The software models internally rate range $10^{-6}/\text{s} < \dot{\gamma} < 10^{6}/\text{s}$ or even wider. Many times with polymers viscosity starts to decrease again at very low rates below $10^{-3}/\text{s}$. Thus this classical zero viscosity is a little bit of a haze definition. The limit for detecting structure components is coming from sensitiveness of rheometer. The rate steps must be as small as possible and the measuring error for the relation $\log \eta / \log \dot{\gamma}$ or respective between data points is the mandatory matter. But needed time to get steady state data point is surely shorter than with polymers.

The form of electrophoregram converted from apparent MMD_a may differ from gained electropherogram analyses. But the scale for M_a is now linear for distribution $w(\log M_a)$. The principle and software employed have no limitations in detecting more sensitive structures; however, the limits are imposed by the accuracy of rheometers in viscosity measurements.

15.8 Conclusions

A new procedure for characterizing blood structures is presented herein, based on a recently developed Unified Model. This computational hemorheological method is introduced using the terminology of polymer rheology and a new linear viscoelastic principle grounded in control theory. While the principle underlying our procedure is straightforward, the numerically sensitive and labile recursive exponential formulas require accurate data and specialized software to achieve high computational accuracy. The computed rheologically effective distribution (RED) serves as an accurate linear fingerprint of the structure, as well as the rheological and viscoelastic properties. From the RED, the electropherogram and apparent molar mass distribution (MMD_a), relaxation modulus, and spectrum are computed. Further developments are discussed. In conclusion, this computational hemorheological method represents a versatile and novel tool for describing the flow properties and analysing blood and its components and fractions.

References

1. Brust, M., Schaefer, C., Doerr, R., Pan, L., Garcia, M., Arratia, P. E., & Wagner, C. (2013). Rheology of human blood plasma: Viscoelastic versus newtonian behavior". *Physical Review Letters, 110*, 078305.
2. Jaishankar, A., Sharma, V., & McKinley, G. H. (2011). Interfacial viscoelasticity, yielding and creep ringing of globular protein–surfactant mixtures. *Soft Matter, 7*(17), 7623–7634.
3. Copley, A. L., King, R. G. (1972). Viscous resistance of thromboid (thrombus-like) surface layers in systems of plasma proteins including fibrinogen. *Thrombosis research 1*(1).
4. Sousa, P. C., Pinho, F. T., Oliveira, M. S. N., & Alves, M. A. (2011). High performance microfluidic rectifiers for viscoelastic fluid flow. *Biomicrofluidics, 5*, 014108.
5. Gross, B. (1953). *Mathematical structure of the theories of viscoelasticity*. Hermann & Cie.
6. Tschoegl, N. W. (1989). *The phenomenological theory of linear viscoelastic behavior: An introduction*. Springler-Verlag.
7. Ferry, J. D. (1980). *Viscoelastic properties of polymers*. Wiley.
8. Sousa, P. C., Pinho, F. T., Alves, M. A., & Oliveira, M. S. N. (2016). A review of hemorheology: Measuring techniques and recent advances. *Korea-Australia Rheology Journal, 28*, 1–22.
9. Alves, M. M., Rocha, C., & Gonçalves, M. P. (2013). Study of the rheological behaviour of human blood using a controlled stress rheometer. *Clinical Hemorheology and Microcirculation, 53*, 369–386.
10. Cherry, E. M., Eaton, J. K. (2013). Shear thinning effects on blood flow in straight and curved tubes. *Physics of Fluids 25*, 073104.
11. Cho, Y. I., Cho, D. J., & Rosenson, R. S. (2014). Endothelial Shear Stress and Blood Viscosity in Peripheral Arterial Disease. *Current Atherosclerosis Reports, 16*, 404.
12. Sloop, G. D., De Mast, Q., Pop, G., et al. (2020). The Role of Blood Viscosity in Infectious Diseases. *Cureus, 12*(2), 7090.
13. Fåhræus, R. (1929). The suspension stability of the blood. *Physiological Reviews, 9*, 241–274.
14. Fåhræus, R., & Lindqvist, T. (1931). The viscosity of the blood in narrow capillary tubes. *American Journal of Physiology, 96*, 562–568.
15. Kensey, K. R., Cho, Y. I., & Chang, M. (1997). Effects of whole blood viscosity on atherogenesis. *The Journal of Invasive Cardiology, 9*, 17–24.

16. Cho, Y. I., & Kensey, K. R. (1991). Effects of the non-Newtonian viscosity of blood on hemodynamics of diseased arterial flows: Part 1, steady flows. *Biorheology, 28*, 241–262.
17. Kim, Y. H., Van de Vord, P. J., & Lee, J. S. (2008). Multiphase non-Newtonian effects on pulsatile hemodynamics in a coronary artery. *International Journal of Numerical Methods in Fluids, 58*, 803–825.
18. Yilmaz, F., & Gundogdu, M. Y. (2008). A critical review on blood flow in large arteries; relevance to blood rheology, viscosity models, and physiologic conditions. *Korea-Australia Rheology Journal, 20*, 197–211.
19. Laun, M., Auhl, D., Brummer, R., Dijkstra, D. J., Gabriel, C., Mangnus, M. A., Handge, U. A. (2014). Guidelines for checking performance and verifying accuracy of rotational rheometers: viscosity measurements in steady and oscillatory shear (IUPAC Technical Report). *Pure and Applied Chemistry 86*, 1945–1968.

Chapter 16
Comparison with Other Known Models

Differences and similarities with well-known models are discussed. At first is one illustration to explain CT principle by Maxwell model and scale arrangement[1]. To compare with Maxwell model was an artificial attempt. In the same publication is good chapters and figures used here elsewhere. Viscoelastic constitutive equations from control theory is derived purely from the standpoint of polymer structure, even though the model was originally developed for detecting the MWD. Microlevel structures and long-chain branches from viscoelastic measurements are discussed earlies.

16.1 Lever Scale for Maxwell Model

We illustrate the familiar Maxwell model in Fig. 16.1a using a modified lever scale comprising series spring-and-dashpot elements on the left side, where each dashpot represents viscosity η_i and each spring represents rigidity G_i. Relaxation time λ_i of the element is defined as $\lambda_i = \frac{\eta_i}{G_i}$, and is a measure of the time required for stress relaxation. The right side includes relaxation-moduli elements presented in the schema shown in Fig. 16.1a by loads as a function of time steps t_i. Since the lever is always in balance, we can neglect the distance to the pivot point of the bar and assume that the model is linear in the mathematical sense. The Maxwell model for an arrangement of mechanical elements represents a discrete spectrum of relaxation times for relaxation modulus $G(t) = \sum_{i=1}^{i=N} g_i \, \mathrm{e}^{-\frac{t}{\lambda_i}}$, where g_i is a Maxwell element. Adding an infinite number of elements will produce the continuous formula

$$G(t) = \int_{-\infty}^{\infty} H(\lambda) \, \mathrm{e}^{-\frac{t}{\lambda}} \, \mathrm{d} \ln \lambda \tag{16.1}$$

T. Borg, *Control Theory in Rheology*, https://doi.org/10.1007/978-3-031-88248-7_16

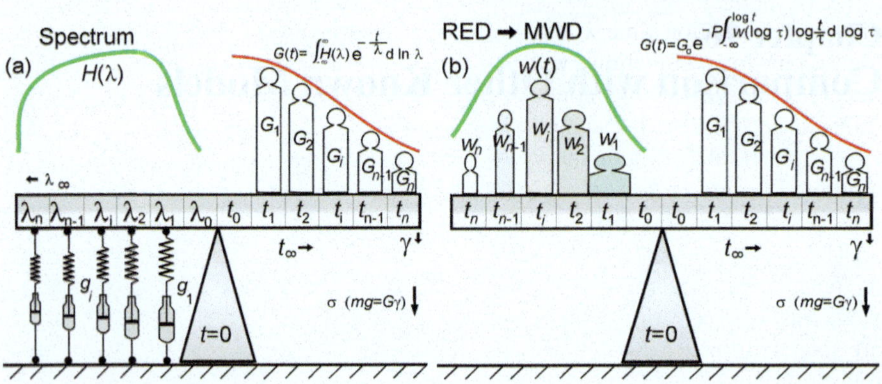

Fig. 16.1 Schematic comparison of the mechanical models using lever scales with dissimilar-type integrals for relaxation modulus $G(t)$ as a function of relaxation time λ and time t. The strongest element g_1 or widest load w_1 relate to the longest and most effective chains. **a** Maxwell model of spring-and-dashpot arrangements for weight loads illustrating $G(t)$. **b** New model based on the rheologically effective distribution (RED), which is related to the molecular-weight distribution (MWD) and the structure of the polymer. Borg and Pääkkönen [1]

where H(λ) is a relaxation spectrum. This much-used formula is mathematically valid and is descriptive for introducing viscoelasticity. Unfortunately, the formulation principle means that this model cannot accurately account for the various non-linear viscoelastic behaviours of real polymers to model and Eq. (16.1) influenced by long molecules, since the orientation and amount of chain entanglements vary with the flow state. Although the spectrum itself is an elegant tool, it has limitations when considering the flows of complex microstructures.

On the lever scale shown in Fig. 16.1b by CT and Eq. (16.2).

16.2 Differences Between Models

The classical Boltzmann superposition principle is thought to hold for any combination of small strains, which are linearly additive. New constitutive Eq. (16.2) fulfils the above condition and moreover has time-strain separability like the Maxwell model. Large strain and shear values are represented using the revised Boltzmann superposition principle in the new time-rate separability model discussed in the Chap. 10. We can develop a model for the lever scale shown in Fig. 16.1b by converting Eq. (6. 5) by CT into the exponential form:

$$G(t) = G_0 e^{-P \int_{-\infty} \log t w(\log \tau) \log \frac{t}{\tau} d \log \tau} \tag{16.2}$$

Comparison of control theory (in this equation) with the Maxwell model in Eq. 16.1 shows that there are different type limits for integrals in addition to the nature of relaxation time λ and Maxwellian exponential $\exp(-t/\tau)$ with the new logarithmic $\log t/\tau$ functions, where τ is the standard time variable in the convolution integral. When describing monodisperse materials and especially polydisperse materials it is necessary to use some kind of distribution function as relaxation spectrum $H(\lambda)$ or standard distribution function $w(\log t)$ RED for modelling viscoelasticity. In a pure relaxation procedure the $w(\log t)$ RED could be called a kind of relaxation spectrum function, but this may conduct to difficulties with shear induced flows.

This apparent similarity of relaxation time λ and standard time variable t suggests having mathematical relation of the models. The single relaxation time is defined $\lambda_i = \eta_i/G_i$, but the time variable t is a normal linear scale in our model. The base variables are totally different and not directly related to each other. In the case modelled relaxation modulus $G(t)$ identical by both models it is the only common property. We illustrate schematic models of the relaxation modulus and the form of REDs, $w(\log t)$ distributions. Both Analytical and Unified Model for relaxation modulus $G(t)$ fits accurately with Maxwellian single-element relaxation in Fig. 16.2a.

Figure 14.2b shows respective Analytical and Unified REDs, $w(\log t)$ and normal (Gaussian) distribution for comparison. On the other hand we have shown, that both of proposed formulas can model most polymer flows. This means that by control theory is possible to model other models and true polymer flows. In conclusion, the only known mathematical relation between Maxwell and the proposed model can be common in the special case that $G(t)$ functions are identical. At this demonstration level the proposed models look more complicated, but they are still similar for more complicated flows.

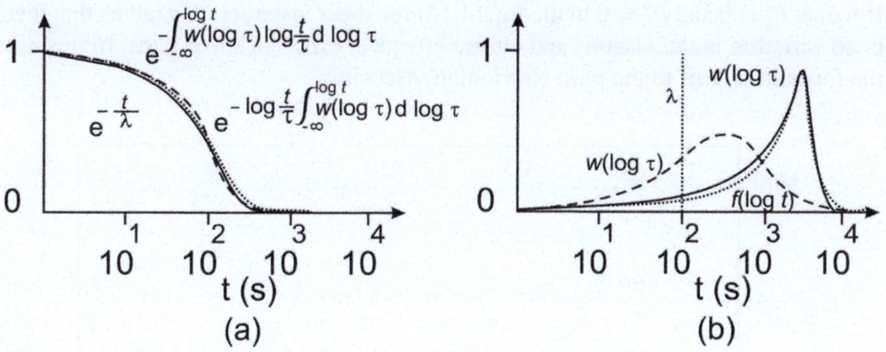

Fig. 16.2 Schematic models of the relaxation modulus and rheologically effective distribution (RED). **a** Unified relaxation modulus plotted on a logarithmic scale with $\tau_c = 1$ s is very close to the classical Maxwellian single-element relaxation, with $\lambda = 100$ s (dotted line) and Analytical relaxation modulus (dashed line). **b** Computed rheologically effective distribution (RED), $w(t)$ from relaxation function. Analytical RED (dashed line) has different melt calibration constants in converting to the close same final MWD form. Only for comparison is added normal (Gaussian) distribution $f(\log t)$ (dotted line) and single-element relaxation bar λ. Borg and Pääkkönen [1]

16.3 Newtonian Fluids and Water at Different Flows

Newtonian fluids represent the simplest mathematical models of fluids that account for viscosity. Water, often used as an example, exhibits pure Newtonian behaviour under ambient conditions. We model the viscosity of pure water using the Unified Model and briefly discuss some possible variations in properties under special circumstances. Due to the low viscosity, forces leading to potential transformations are minimal for viscosity values.

For supercooled water [2], strong shear rate thinning effects at high shear rates have been observed, similar to those seen with standard polymers. Temperature dependency also has a strong effect on water's viscosity values. Measurements of low-viscosity fluids at high shear rates with standard rheometers face several challenges, such as slippage between solid surfaces and liquids. Nevertheless, it has been observed that Newtonian liquids might exhibit shear thinning when shear rates exceed the range of 10^7/s to 10^8/s. Water at ambient temperature with standard oscillation rheometer is measured very slight shear thinning property at rates 10^5/s to 10^6/s, but one possible reason for that is explained [3, 4] slippage between solid surfaces and liquids.

In recent years, developments in microfluidic rheology have been achieved, enabling viscosity measurements and modelling at extremely high shear rates[5].

The obtained shear thinning effect in Fig. 16.3 can be true coming from molecules or instrumental and system errors.

We model similar way viscosity flow curve for pure water with slight shear thinning effect simulating measurements as in earlier Sect. 4.14 for steady shear viscosity. If we set $P' = 0$ and $P'' = 0$ in the Eq. (4.16) for shear viscosity, this tell us that there is no variation in the viscous and elastic effects at different shear rates. In this case the formula returns to the pure Newtonian viscosity.

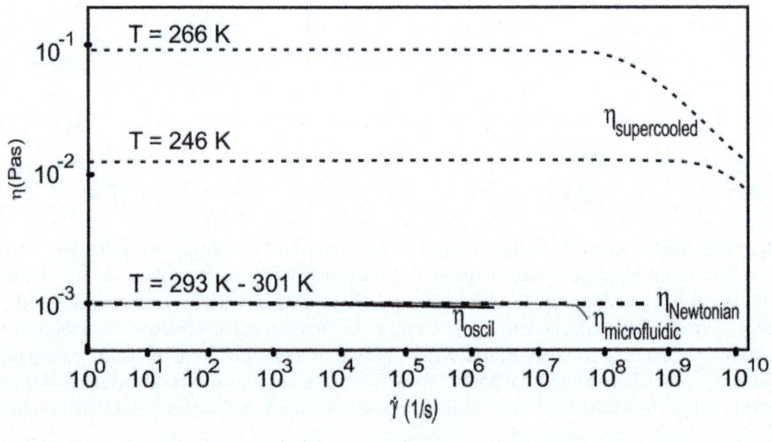

Fig. 16.3 Composition of obtained or modelled viscosity of pure water by different measurements and temperatures

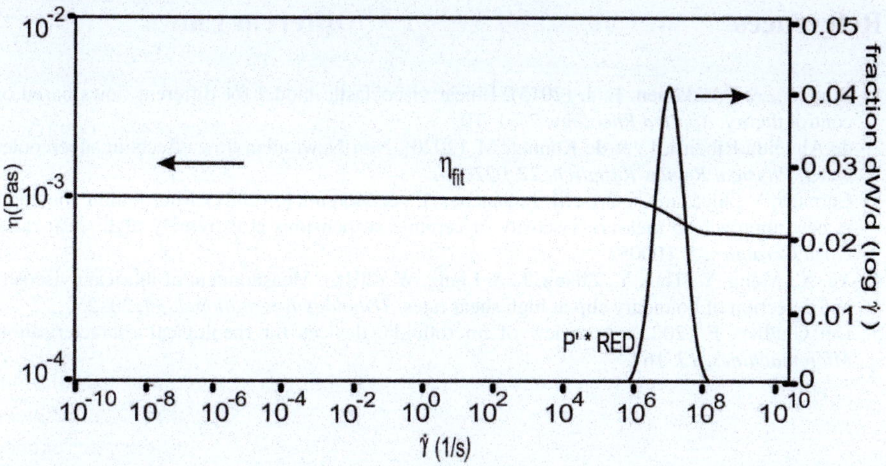

Fig. 16.4 Composition of viscosity for pure water at different shear rates. This figure models the viscosity of pure water at ambient temperature across various shear rates. The Wesslau distribution was applied using parameters $\mu = 7$ and $\sigma = 0.4$. Additionally, we set $P' = 0.04$, $P'' = 0$, and log $R = 15$ in the spreadsheet depicted in Fig. 4.8

In the Fig. 16.4 is shown modelled viscosity curve for water at ambient temperatures in range 20–28 °C. Water is thought pure viscous fluid and we set $P' = 0$ to the spreadsheet in Fig. 4.8. As we do not know true origin of constant P', we do not continue to analyse the results.

16.4 Summary

In summary, the proposed model may at the first sight appear to be Maxwellian, but it is actually fundamentally different due to control-theory principles and missing relaxation time and spectrum schemas. In addition, the variables and integrals have completely different types of structure. By our model is possible to simulate relaxation modulus modelled by the Maxwell model or true measured $G(t)$ of any polymers, but on the contrary mostly impossible. Although polymer behaves nearly never linearly on the standpoint of Newtonian or pure Maxwell model even with vast amount of elements, modelling is possible by control theory inside measuring errors according to our experiences. Moreover, used formulas are always linear as the control theory is known a mathematically linear principle.

References

1. Borg, T., & Pääkkönen, E. J. (2015). Linear viscoelastic model for different flows based on control theory. *Applied Rheology, 25*, 1–10.
2. de Almeida Ribeiro, I., & de Koning, M. (2020). Non-Newtonian flow effects in supercooled water. *Physical Review Research, 22*, 022004.
3. Carnicer, V., Alcázar, C., Orts, M. J., Sánchez, E., & Moreno, R. (2021). Microfluidic rheology: A new approach to measure viscosity of ceramic suspensions at extremely high shear rates. *Open Ceramics, 5*, 100052.
4. Yu, X., Meng, Y., Tian, Y., Zhang, J., & Liang, W. (2016). Measurement of lubricant viscosity and detection of boundary slip at high shear rates. *Tribology International*, *94*, 20–25.
5. Del Giudice, F. (2022). A review of microfluidic devices for rheological characterisation. *Micromachines, 13*, 167.

Glossary of Symbols

Symbols

a	Sampling band width
a_{T}	Horizontal TTS constant
b_{T}	Vertical TTS constant
a'_T	Temperature coefficient for elastic value
a''_T	Temperature coefficient for viscosity value
D	Dimension exponent
D_s	Diameter of the tube
F_{c}	Elastic force for a single chain
f_{i}	Gaussian value
F'	Elastic connector force
G^*_∞	Limit complex modulus
$G_{\mathrm{c}}(t, T)$	Relaxation modulus by time and temperature
G^0_{N}	Plateau modulus
G'	Dynamic storage modulus
G''	Dynamic loss modulus
$\mathrm{g}(t)$	Normalized relaxation modulus
$G(t)$	Relaxation modulus
G^*	Complex modulus
G_0	Relaxation modulus at $t < 10^{-4}$ s
$G(t, \gamma)$	Imposed-ramp relaxation function
$h(t)$	Impulse response
$h(\lambda)$	Relaxation spectrum
Hf	Conversion factor between scales
k	Flow consistency index by Ostwald-de-Waele
K	Spring constant
L_s	Statistical unit segment length
Mf_0	Polymer structural value at reference temperature T_0
$\overline{M}_W$	Weight average molecular weight

T. Borg, *Control Theory in Rheology*, https://doi.org/10.1007/978-3-031-88248-7

Mw	Average molecular weight
MwR	Polydispersity index
M	Molecular weight
$M(t_0)$	Average molecular size
M_0	Molecular weight of the monomer
Mf	Polymer structural value
M_i	Molecular weight M of each fraction
MWD, $w(\log M)$	Molecular weight distribution
MWD, $w_{\mathrm{MAIN}}(M)$	Molecular weight distribution by GPC/SEC
n	Power law index
N_A	Avogadro's number
$\widehat{P}$	Effective polymer structure function
P'	Elastic constant, entanglement value
P''	Viscosity constant
$P(\dot{\gamma}), P(\omega)$	Polymer structure function
P_H	Maximal strain-hardening constant
$P_H\dot{\varepsilon}$	Strain-hardening variable
$\widehat{P}_{REL}\%$	Percentage effective polymer structure function
$P(T)$, P^T	Polymer structure value as a function of temperature
R	Rouse constant
%RMSE	Percentage root-mean-square error function
$\langle R^2\rangle_0$	Mean-square end-to-end distance
$r(t)$	Radius of sphere
t	Time
t_0	Time at t=0
t_c	Characteristic time
V_0	Nonrelaxed sphere of volume
$w(\log t)$	Rheologically effective distribution (elastic)
$w_\mathrm{H}(\log\ t, \dot{\varepsilon})$	Strain-hardening distribution by time and rate
$\check{w}(\log\dot{\gamma}), \check{w}(\log\omega)$	Normalized rheologically effective distribution RED
$\check{w}(\log M)$	Normalized molecular weight distribution MWD
$w_\mathrm{H}(\log\ \varepsilon)$	Original strain-hardening component
$w(\dot{\gamma})$	RED as a function of shear rate $\dot{\gamma}$
$w_{\mathrm{LCB}}(M)$	Long-chain branching distribution
$w'(\log\dot{\gamma})$	Effective elastic distribution RED'
$w''(\log\dot{\gamma})$	Effective viscous distribution RED"
$w(t)$	RED as a function of time t
$w(\omega)$, $w(\log\omega)$	Rheologically effective distribution RED
w_{GPC}	MWD by GPC
w_i	Molecular weight fraction
$w(\log T)$	Rheologically effective distribution for temperature
$\dot{x}$	General rate
$y(t)$	Pulse response

Greek Letters

$\dot{\gamma}$	Shear rate
$\widehat{\dot{\gamma}}_0$	Effective zero shear rate
$\dot{\gamma}_{\mathrm{c}}$	Characteristic shear rate, normally $\dot{\gamma}_{\mathrm{c}} = 1/\mathrm{s}$
ε	Hencky strain
$\dot{\varepsilon}$	Elongation rate, Hencky strain rate
$\hat{\eta}_0$, $\hat{\eta}_0^*$, $\hat{G}_0$	Effective zero states depending on the flow type
$\tilde{\eta}_{\mathrm{E}}^+(\mathrm{t})$	Elongational viscosity without strain hardening effect
$\widehat{\eta}$	Old viscosity at $t = t_0$
$\eta(\dot{\gamma}_0)$	Steady-state limit viscosity
η^*	Complex viscosity
$\eta(\log \dot{\gamma})$, $\eta(\dot{\gamma})$, η	Shear viscosity
$\eta * (\omega, T)$	Complex viscosity by frequency and temperature
$\eta^-(t)$	Shear-stress decay coefficient
$\eta^+(t, \dot{\gamma})$	Shear stress growth coefficient
$\eta^*(\omega)$	Complex viscosity
η_0, η_0^*	Zero viscosity
η_0^+, $\tilde{\eta}_0^+$	Elongational viscosity at characteristic time $t_{\mathrm{c}} = 1/\mathrm{s}$
$\eta_0^*(T)$	Zero viscosity as a function of temperature
η_∞^*	Limit viscosity
$\eta_{\mathrm{c}}(\dot{\gamma}, T)$	Shear viscosity by shear rate and temperature
$\eta_{\mathrm{E}}^+(t)$	Elongational viscosity with strain hardening effect
$\eta(t, \dot{\gamma})$	Time-dependent transient shear viscosity
η_E	Trouton viscosity
λ	Maxwellian single-element relaxation time
μ	Mean in distribution
ξ	Friction coefficient
σ	Standard deviation in distribution
σ_{E}	Net tensile stress monitoring as a function of time
σ^+	Stress growth function
$\sigma^+(t, \dot{\gamma})$	Shear-stress growth coefficient
σ_t^+	Time-dependent relaxation stress
σ_ε^+	Elongation stress of the shear
τ_{c}	Characteristic time $\tau_{\mathrm{c}} < 10^{-4}$ s
$\chi(t, \dot{\varepsilon})$	Strain-hardening coefficient
ψ, τ	Integration variable for rate $\dot{\gamma}$, frequency ω or time t
ω	Frequency
ω_∞	Critical limit rate

Author Index

T. Borg, *Control Theory in Rheology*, https://doi.org/10.1007/978-3-031-88248-7

Subject Index

T. Borg, *Control Theory in Rheology*, https://doi.org/10.1007/978-3-031-88248-7

The manufacturer's authorised representative in the EU is Springer Nature Customer Service Centre GmbH, Europaplatz 3, 69115 Heidelberg, Germany. If you have any concerns regarding our products, please contact ProductSafety@springernature.com

Printed and bound by CPI Group (UK) Ltd, Croydon, CR0 4YY
30/09/2026
02225586-0003